T0135670

Inductively Coupled Microsensor Networks

Relay Enabled Cooperative Communication and Localization

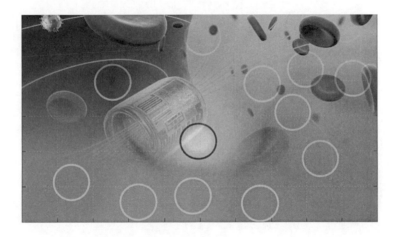

Eric Slottke

Series in Wireless Communications
edited by:
Prof. Dr. Armin Wittneben
Eidgenössische Technische Hochschule
Institut für Kommunikationstechnik
Sternwartstr. 7
CH-8092 Zürich

E-Mail: wittneben@nari.ee.ethz.ch
Url: http://www.nari.ee.ethz.ch/

Bibliographic information published by the Deutsche Nationalbibliothek

The Deutsche Nationalbibliothek lists this publication in the
Deutsche Nationalbibliografie; detailed bibliographic data are
available in the Internet at http://dnb.d-nb.de .

ISBN 978-3-8325-4438-6
ISSN 1611-2970

Logos Verlag Berlin GmbH
Comeniushof, Gubener Str. 47,
10243 Berlin
Tel.: +49 030 42 85 10 90
Fax: +49 030 42 85 10 92
INTERNET: http://www.logos-verlag.de

Diss. ETH No. 23737

Inductively Coupled Microsensor Networks

Relay Enabled Cooperative Communication and Localization

A thesis submitted to attain the degree of
DOCTOR OF SCIENCES of ETH ZURICH
(Dr. sc. ETH Zurich)

presented by

ERIC NATHAN SLOTTKE

Dipl.-Ing., Leibniz Universität Hannover
born on July 07, 1985
citizen of Germany and the United States of America

accepted on the recommendation of
Prof. Dr. Armin Wittneben, examiner
Prof. Dr. Christoph Mecklenbräuker, co-examiner

2016

Day of Doctoral Examination: December 12, 2016

Abstract

In this thesis, we study a novel paradigm for wireless sensor networks: we envision a dense *microsensor network*, consisting of hundreds or thousands of highly miniaturized wireless nodes with millimeter or sub-millimeter dimensions. Such a microsensor network has many interesting applications ranging from in vivo medical sensing to environmental monitoring. However, the design and operation of the envisioned type of network are challenging: the large number of nodes, in combination with the small form factor, imposes severe constraints on both node complexity as well as power consumption. We propose using inductive near-field coupling as advantageous physical layer choice, allowing an RFID-like operation of the network with wireless power supply from central reader devices and low-complexity tag design. The use of inductive near-field coupling has only rarely been studied in the context of microsensor networks.

Our primary goals for inductively coupled microsensors are twofold: we want to enable reliable communication to and between sensor nodes, and perform accurate localization of individual sensors. Both tasks are affected by the central limitation of a physical layer based on near-field coupling: the severely limited range of interaction. We will show throughout this thesis that the use of wireless relaying allows for overcoming this limitation. Based on a circuit-theoretic communication framework extended to inductively coupled microsensor networks, we first explore fundamental properties and design limitations. We then investigate suitable approaches to relaying and study their use to increase both the communication range and link reliability.

In addition to improving the communication performance of microsensor networks, we demonstrate the capability of inductively coupled relays to enable novel secondary uses of the wireless channel. To this end we show that relay-assisted microsensor networks can achieve distributed computation by introducing a scheme to implement wireless artificial neural networks over multihop MIMO channels. We continue by showing the feasibility of accurate localization in an RFID-like setting. To this end, we propose a novel method that allows for the localization of a purely passive sensor only consisting of a simple loop antenna and a matching circuit. The unknown position of the sensor is hereby reconstructed from measuring the input impedance at several reference nodes. By the use of relaying, this scheme can also be adopted to networks with only a single measurement entity. To verify the obtained theoretical and simulative results, we present an experimental setup and perform a practical evaluation of selected schemes. In this context, appropriate methods for mitigating the imperfections of real-world measurements are developed.

Kurzfassung

Dieser Dissertation liegt ein neuartiges Konzept für drahtlose Sensornetzwerke zugrunde: dichte Netzwerke, bestehend aus hunderten oder tausenden von hoch miniaturisierten drahtlosen Knoten mit Abmessungen im Millimeter- oder Submillimeter-Bereich. Diese von uns als *Mikrosensornetzwerk* bezeichnete Konfiguration ermöglicht eine Vielzahl interessanter Anwedungen von medizinischer Sensorik im Körperinneren bis hin zur Erfassung kleinräumiger Umweltphänomene. Die Anzahl der Funkknoten sowie deren Miniaturisierung führt hierbei zu Herausforderungen etwa in Fragen der Energieversorgung oder der Systemkomplexität einzelner Funkknoten. Wir schlagen in diesem Zusammenhang die Verwendung von induktiver Nahfeldkopplung als Übertragungstechnologie für Mikrosensornetzwerke vor. Diese Technologie erlaubt eine externe Energieversorung passiver Funkknoten mit geringer Komplexität. Die Verwendung induktiver Kopplung als Übertragungstechnologie für Mikrosensornetzwerke wurde bislang in der Literatur kaum behandelt.

Im Kontext induktiv gekoppelter Mikrosensornetzwerke verfolgen wir zwei vorrangige Ziele: zum einen die zuverlässige Kommunikation mit und zwischen einzelnen Funkknoten, sowie die genaue drahtlose Ortung einzelner Knoten innerhalb des Netzwerks. Hierbei stellt die Beschränkung der Interaktion induktiv gekoppelter Funkknoten auf den Nahfeldbereich für beide Aufgaben eine zentrale Herausforderung dar. Wir zeigen dass diese Problematik durch den Einsatz einzelner Funkknoten als Relays bewältigt werden kann. Auf Basis eines an induktiv gekoppelte Mikrosensornetzwerke angepassten, schaltungstheoretischen Kommunikationsmodells betrachten wir zunächst fundamentale Eigenschaften und Grenzen solcher Netzwerke. Wir untersuchen weiterhin wie durch den Einsatz induktiv gekoppelter Relays die Kommunikationsreichweite und Verbindungsqualität verbessert werden können.

Neben der Unterstützung der Kommunikation in Mikrosensornetzwerken demonstrieren wir neuartige sekundäre Nutzungsmöglichkeiten des drahtlosen Kanals, welche durch induktiv gekoppelter Relays ermöglicht werden. Hierzu zeigen wir dass Mikrosensornetzwerke mithilfe von Relays verteilte Berechnungen durchführen können, indem wir ein Verfahren zur Implementierung drahtloser künstlicher neuronaler Netzwerke über Multihop-MIMO-Kanäle einführen. Weiterhin demonstrieren wir die Möglichkeit der genauen Funkortung in induktiv gekoppelten Systemen. Wir schlagen hierzu eine neuartige Methode zur Ortung eines komplett passiven Sensorknotens vor, welcher lediglich aus einer einfachen magnetischen Antenne und einem passiven Anpassnetzwerk besteht. Die unbekannte Position des Sensors wird hierbei aus Messungen der Eingangsimpedanz an mehreren Referenzknoten rekonstruiert. Durch die Verwendung von Relays kann diese Methode auch auf Netzwerke mit einer einzelnen Messeinheit erweitert werden. Die erhaltenen theoretischen und simulativen Ergebnisse für die Funkortung in induktiv gekoppelten Systemen werden anhand eines Messaufbaus validiert. In diesem Kontext werden geeignete Methoden zum Umgang mit Imperfektionen in praktischen Messungen entwickelt.

Contents

Part I.

Introduction

1

Inductively Coupled Microsensor Networks

Throughout the last decades, one of the main driving forces behind innovation in electronic systems has been the process of ongoing miniaturization. Since the development of integrated circuits, the number of components per chip area has doubled roughly every one to two years, an observation prominently known as Moore's law [101]. This rapid growth has lead to the emergence of devices such as modern cellular handhelds or desktop computers.

Specifically in the context of wireless communication, there has been an interest to not only miniaturize individual hardware components, but to manufacture integrated, ultra-small devices with the capability to communicate among themselves as well as to sense or manipulate their environment [1], [3]. We envision a class of wireless networks consisting of a large number, potentially hundreds or thousands, of highly miniaturized nodes with sizes in the sub-millimeter range. Similar to the notion of a classical macroscale sensor network [115], [5] with device dimensions in the centimeter scale or larger, these nodes have the ability to sense and interact with their environment. We refer to this type of network as *microsensor network*.

1.1   The Paradigm of Microsensor Networks

Microsensor networks are envisioned to be employed in settings where macroscale sensors are too invasive or too large to interact with small, localized phenomena. A field of high relevance for potential applications of microsensor networks is found in medicine. Due to the small size of the individual nodes, they can perform various tasks inside the human body. A well-known example of an in-body sensing application is wireless capsule endoscopy [65], in which a node equipped with a miniature camera wirelessly transmits image data from the gastrointestinal tract of a patient. The typical size of an endoscopic capsule is on the order of several millimeters. At smaller scales, wireless nodes are envisioned for targeted drug delivery [30] or localized in-body sensing of oxygen supply [38]. A large collection of further potential applications for microsensor networks in the medical field can be found in [106] and [138].

Several applications of microsensor networks have also been proposed outside the medical field. An interesting use lies in the field of engineering, where micro- and nanoelectronic devices are envisioned to act as building blocks for matter with programmable properties [46]. Such networks are envisioned to self-assemble into clusters of nodes which are reconfigurable both in shape and in function, e.g. for use in design and prototyping.

1.1.1 Requirements and Challenges

Considering the envisioned applications of microsensor networks, we identify a set of requirements governing their design from an engineering standpoint. Such an analysis requires a set of underlying assumptions on the general properties of microsensor networks which is partly speculative in nature. Basic requirements, trade-offs, and limitations, however, can partly be inferred from the physical size of the nodes and the intended applications.

Communication A central requirement for the envisioned applications of microsensor networks lies in the ability of the individual nodes of the network to communicate in order to extract collected sensor data from the network, receive commands, or perform cooperation between the nodes. Communication can be required both among individual nodes as well as between a node and a central unit coordinating the network. The communication should be wireless to enable networking for ad-hoc node deployment. There exist several fundamental limitations on the communication performance of size-limited nodes using radio-frequency (RF) electromagnetic signaling. On one hand, the radiation efficiency and resonance bandwidth of electrically small antennas are inherently decreased with antenna size [171]. On the other hand, the limited energy storage for transmission leads to a trade-off between the communication performance and energy autonomy of the nodes.

Power supply Microsensor nodes require power both for the operation of the node as a whole, as well as specifically for communication. Three different approaches exist to supply power to untethered devices: local energy storage, external power supply, and energy harvesting. Macroscale sensor networks typically store energy locally, e.g. in chemical form using batteries. While this method imposes no requirements on the environment of the network, the amount of locally stored energy and therefore the time of unserviced operation is finite, with the amount of stored energy typically being constrained by the volume of the storage device. The authors of [178] investigate the use of capacitors, supercapacitors, galvanic cells, fuel cells, and radioisotopic energy converters as sub-millimeter sized local energy storage mechanisms. Herein galvanic cells are identified as providing the best combination of energy storage and power supply.

It should be noted that manual servicing of microsensor nodes, e.g. to recharge or replace a local energy storage device, is prohibitive for many applications such as in-body data sensing. Relying on local energy storage in small-scale wireless nodes therefore results in the increasingly important design requirement of low energy consumption as the node size is decreased [23]. Particularly for microsensor networks, external power supply and energy harvesting are viable alternatives to ensure a sustained, maintenance-free operation of the network. In energy harvesting, energy is continually extracted from sources present in the environment of the nodes. This requires the nodes to convert the energy into a form that can be stored locally. Energy harvesting in small-scale systems has been proposed and demonstrated for a variety of energy forms, including energy from light sources, thermal sources, electromagnetic radiation, or vibrations [64]. However, depending on the environment, the power that can be extracted from energy harvesting mechanisms may be below the amount required for continuous operation of the network, possibly limiting the applicability of energy harvesting. On the other hand energy can also be made available artificially for the nodes to convert, a process referred to as external wireless power supply. External power supply in sensor networks is realized most commonly in the form of electromagnetic energy, i.e. either using propagating waves or near-field coupling [126]. Both energy harvesting and external power supply enable microsensor nodes to operate autonomously over a long time.

Complexity We can expect microsensor nodes to generally be of lower complexity than their macroscopic counterparts. In this context, we use the word complexity as a broad term encompassing computational power, communication resources, and number of sensing and actuation functions. The reasons for its restriction are twofold: first, integration of higher complexity increases power consumption, which is subject to the above mentioned limitations. Furthermore the overall size of the node itself constrains the amount of functionality that can be integrated.

Localization A key requirement for several of the envisioned applications of microsensor networks is the ability to perform localization, i.e. to locate one or multiple nodes either in an absolute coordinate system or relative to the other nodes in the network. Specific applications might require the data produced by the sensor nodes to be associated with their respective positions. An example is found in wireless capsule endoscopy where recorded video data can be interpreted more easily if it can be matched to the position of the capsule inside the body. The relative position between several nodes can also be of interest e.g. for cooperative actuation tasks.

Several approaches exist to perform wireless localization (cf. Chapter 6), which all have in common that the produced location estimates generally include random errors with magnitudes depending on several parameters such as signaling bandwidth or signal-to-noise ratio. Considering these errors, an interesting aspect of localization in microsensor networks is given by the overall dimensions of the network: If the network is reduced to a size such that the typical error of a location estimate is significantly larger than the typical distance between the microsensor nodes, relative arrangements of multiple nodes can not be reliably reconstructed from localization. Accordingly, the specific choice of localization scheme is constrained primarily by the accuracy requirements of the intended applications in addition to the complexity limits of the nodes.

1.1.2   Physical Layer Technologies

Many of the performance characteristics of a wireless communication system are a consequence of the employed physical layer technology. With the envisioned paradigm of microsensor networks still being primarily conceptual in nature, there is no standardized physical layer in existance for this type of networks. However, a range of suitable technologies have been proposed for communication with highly miniaturized devices, each having distinct benefits and drawbacks. In the following we provide a brief overview of physical layer technologies suggested for the communication among microsensor nodes.

RF Electromagnetics The utilization of a physical layer based on RF electromagnetics has the inherent benefit of drawing on well-understood theory and a large body of standardization. Particularly the IEEE 802.15.4 standard is widely used for wireless sensor networks and body area networks on the macroscopic scale. Its applicability to dense microsensor networks is investigated in [17] in the context of energy efficiency. Another interesting option is the use of ultra-wideband (UWB) impulse radio, which allows for low-complexity, low-power transceiver implementations [152], [141]. UWB allows high peak data rates, making it a sensible choice for the uplink in a microsensor setting [45]. An UWB physical layer can also be employed synergistically to perform accurate localization due to the large signal bandwidth used [99].

Optical Communication Another option proposed for communication in microsensor networks is the use of free space optical microelectromechanical systems (MEMS) as communication link. In free space optical communication, information is generally transmitted using a highly directive light beam. Two approaches exist to incorporate this communication paradigm into microsensor nodes [71]: The node may act as an active transmitter by means of a light source—such as a laser diode—incorporated into the device. On the other hand the power consumption of the node can be reduced by using passive communication based on reflection. To this end, the nodes use steerable mirrors to modulate the reflection on an impinging beam [179]. The benefits of free space optical communication include high power efficiency, high gain, and immunity to multipath induced fading [167]. It should be noted that effective communication requires a line-of-sight connection between the nodes. Furthermore the necessary mechanical steering of both active light sources and reflecting mirrors imposes the requirement of accurate mutual location knowledge of the communicating nodes.

Nanoscale Electromagnetics It is well understood that in the miniaturization of RF wireless communication devices with implementations based on silicon, a point will eventually be reached at which further downscaling becomes infeasible due to physical effects dominating the device behavior at very small sizes [74]. An interesting proposal to further push the boundaries of miniaturization—towards integrated microsensors with sizes in the micrometer regime—is to make use of nanotechnology. As an example, there has been interest in manufacturing antennas out of novel materials such as graphene or carbon nanotubes (CNT) [51], [158]. Due to so-called plasmonic wave propagation experienced in these materials, the resonance frequency of such antennas is found at significantly lower frequencies than for metallic antennas of comparable size. A half-wave dipole measuring $20\,\mu$m fabricated either from perfectly conducting metal or a CNT exhibits a resonance frequency of 7.5 THz and 160 GHz, respectively, as calculated in [51]. This effect facilitates the miniaturization of what is often one of the largest components in integrated microsensors, the antenna.

On the other hand, nanoscale communication systems with operating frequencies well below the THz range can be realized using nanoelectromechanical systems (NEMS). Proof-of-concept NEMS for wireless RF communication have been proposed and even realized in a laboratory setting. For example, a simple working principle for a receiver operating at frequencies between 40 MHz to 400 MHz has been presented in [69] along with a functioning prototype. The device consists of a single CNT which is mounted as a cantilever on one of two parallel plate electrodes, to which a bias voltage is applied with respect to the second electrode. For sufficiently high voltages the electric field around the nanotube will lead to field electron emission, an effect described by Fowler and Nordheim in [43]. The effect is particularly pronounced for carbon nanotubes, as the local field strength at the tip of the CNT is very high due to the nanotube geometry. The field electron emission results in a measurable emission current between the cathodes, while the applied bias voltage additionally leads to a high concentration of electrons at the tip of the nanotube, thus creating a localized net charge. The operation principle of the receiver is based on the electromechanical resonance of the charged tip of the CNT excited by impinging electromagnetic waves. As the nanotube vibrates, the variable distance of its tip to the second electrode leads to an emission current which varies as a nonlinear function of the mechanical vibration, inherently implementing an envelope detector demodulation.

Nanomaterials have also been studied for use in components other than antennas such as sensors, actuators, or power supply. A survey of existing and envisioned NEMS components for microsensor networks is found in [4].

Molecular Communication An entirely different communication paradigm for microsensor networks is found in molecular communication, in which information is exchanged by the physical transport of molecules between a source and destination. This paradigm was proposed in 2005 [58], inspired by the communication based on the exchange of molecules encountered in many biological systems.

An overview of the state of the art in molecular communication can be found in [104]. Herein a network using molecular communication is often assumed to consist of highly miniaturized devices on the micrometer and submicrometer scale. A transmitting device does so by releasing molecules into its environment. Several possibilities have been studied to encode information onto the released molecules, such as using distinct types of molecules to represent different messages [35], or representing information in the rate of emission [103]. The presence of the information carrying molecules is detected at a receiver by means of receptors specifically binding to the molecules of interest. Two concepts have been proposed to implement this communication mechanism: the construction of artificial cells [130], and the modification of biological cells [13]. Molecular communication is particularly interesting for microsensor networks embedded in biological systems due to the potential for interaction of the sensors with signaling from biological sources.

From an information theoretical perspective the primary distinction between molecular communication and electromagnetic wireless systems lies in the mode of propagation. As a result of the Brownian motion of fluids, a released molecule moves away from the transmitter following a path that can be modeled as a random walk. If a large number of molecules is emitted, the statistical description of their spatial density over time accordingly follows the laws of diffusion. Major drawbacks of this transport mechanism include slow propagation, the potential for long delays and information loss, and no guarantee of reception in sequential order [36]. Guidance of the information-carrying molecules using transportation structures—again inspired by biological systems—has been proposed to mitigate these drawbacks [102]. This mode of transport is conceptually similar to using wired connections in macroscopic networks.

In this work we study microsensor networks based on an alternative to the technologies reviewed above, namely inductive coupling. As elaborated in the following section, inductive coupling presents a promising technology choice in consideration of the unique requirements of microsensor networks.

1.2 Inductively Coupled Microsensor Networks

We propose to use inductive coupling as the technology basis to implement communication, localization, and external power supply for microsensor networks. Inductive coupling is an effect of mutual interaction between conductor loops in the non-radiative near-field. A pair of wireless nodes couples by means of a time variant magnetic flux, generated by a current in the antenna of a transmitting node and inducing a voltage in the antenna of a receiver. A formalization of this interaction mechanism is given in Chapter 3. The inductively coupled antennas can be interpreted as air-core transformer over which both energy and information can be transferred, with the intensity of the interaction depending on the magnitude of the linked flux. The lack of a transformer core results in the magnetic flux being spread out in space, with only a fraction of it flowing through both antennas. To achieve efficient energy transfer and low signal path loss, inductively coupled nodes therefore typically need to be placed closely together and oriented properly.

An important approach to increase the efficiency of the interaction between two nodes is the utilization of magnetic resonance. By canceling the reactive voltage drop of the antenna circuits using capacitive elements, both the field-generating current as well as the induced voltage can be increased. Compared to the non-resonant case, resonantly coupled nodes can achieve high transfer efficiencies even for settings where the nodes are moderately far away or misaligned [73]: as will be discussed in Section 4.3, the quality factor of resonantly coupled nodes can partially compensate a suboptimal node placement.

The use of inductive coupling as a physical layer for wireless communication is well-known from its widespread use in radio frequency identification (RFID) [42]. RFID utilizes a distinct benefit of inductively coupled communication, namely the ability to create a highly asymmetrical complexity distribution within the network, meaning that most of the complexity required for communication can be integrated in a central unit while the remaining nodes in the network are low-complexity devices. Accordingly, inductively coupled RFID systems typically are built around a central unit called reader which has the primary goal of wirelessly detecting the identity or presence of low complexity nodes called transponders or tags. The low-complexity design of the tags is enabled by a combination of simple, unidirectional communication schemes such as load modulation (cf. Section 3.2) and external power supply by the central unit [21]. An extreme example of this paradigm is found in 1-bit RFID systems in which a central reader device detects the presence of purely passive nodes with no own logic or power supply. Herein the node's presence or lack thereof corresponds to one bit of information. The typical applications of inductively coupled RFID systems include access control, electronic article surveillance, as well as warehouse and livestock tracking. Inductively coupled RFID systems primarily operate in the range of $9 - 135$ kHz or the ISM band[1] at 13.56 MHz. The extension of RFID technology from a paradigm of asymmetric complexity to generic, bidirectional peer-to-peer communication is specified in a set of standards referred to as near-field communication (NFC) [66], [67]. NFC builds on the inductively coupled physical layer in the 13.56 MHz band. Besides peer-to-peer operation, NFC also allows for the implementation of a reader/tag infrastructure as well as the emulation of various smartcard standards [26]. It has seen widespread adoption e.g. in modern smartphone devices for applications such as mobile payment and access control.

Inductive coupling has also been proposed as a physical layer for wireless systems operating in harsh propagation media. Conductive propagation environments, such as water or soil, typically impose strong attenuation on electromagnetic waves. However, the magnetic permeability of these media is very close to the vacuum permeability and therefore magnetic near-field communication systems experience no significant additional attenuation. Consequently inductively coupled communication systems have been studied for underwater point-to-point communication links [148], [31] and sensor networks buried underground [6], [94].

In the context of the envisioned setting of wireless microsensor networks, inductive coupling

---

[1]Industrial, scientific, and medical (ISM) bands are often used by low-power, short-range communication systems due to the possibility of license-free operation.

shows the potential to fulfill several of the requirements addressed in the previous section. In consequence we identify this technology as a viable physical layer for this type of network. Specifically we anticipate inductive coupling to synergistically achieve four distinct tasks:

Inductively coupled communication The primary task realized by inductively coupled sensor nodes is that of communication which, depending on the application, needs to be possible both between a central unit and the microsensor nodes, as well as between individual nodes.

Inductively coupled power supply Resonant inductive links have been identified as a viable means for high-efficiency wireless power transfer [80]. As argued in Section 1.1.1, external power supply is a key enabler for highly miniaturized, autonomous sensor nodes. By reusing the inductively coupled communication link for power transfer, the overall complexity of the nodes is reduced.

Inductively coupled localization The mechanism of inductive coupling may also be reused to locate individual nodes in the network, referred to as agents. The location of the agents is hereby determined with respect to reference nodes, called anchors. In Chapters 6, 7, and 8 we propose and study a localization scheme that allows accurate localization even if the agent is a purely passive device.

Inductively coupled actuation Finally, inductive coupling is beneficially reused to perform actuation tasks in the microsensor network. As an example, by converting the energy of the magnetic field into heat, in-body microsensor nodes can perform highly localized treatments, e.g. externally triggered drug delivery or the damaging of cancer cells by thermal ablation [41]. Magnetic fields have also been proposed as a mechanism of enabling locomotion of microscale devices [62], [25].

However, inductive coupling also has drawbacks. Most notably the amplitude of the magnetic near-field decays at a rate of d^{-3} with distance d from the source. Compared to communication using propagating electromagnetic waves with field components decaying as d^{-1}, the usable range of inductively coupled systems is drastically smaller. Secondly, while a resonant design leads to an increase in signal transmission efficiency it also corresponds to a smaller transmission bandwidth, limiting the data rate that can be transmitted over such a link. Furthermore the resonance frequency of a node may change as a consequence of its coupling to other nodes [21]. Particularly strong coupling can lead to a significant detuning of the circuit. Dense networks of mobile and resonantly designed nodes may therefore require adaptive tuning of the individual circuits.

The potential of utilizing inductively coupled RFID devices to implement macroscopic sensor network applications has been noted in literature [150], [70]. However, many macroscopic sensor network designs consider inductive coupling purely for power supply alongside an RF communication link [175]. For example, the authors of [8] demonstrate an inductively powered wireless sensor network utilizing the ZigBee standard for communication at 2.4 GHz. Due to the limited range of inductive coupling, the power source must be placed close to the sensors in such a setting. To provide power to a large, dispersed RF sensor network, the approaches in [137], [85] propose the use of a mobile vehicle capable of recharging individual sensor nodes via an inductively coupled link. The use of inductive coupling to jointly realize power supply and communication is well established in the medical field, e.g. for the wireless monitoring implanted devices such as cochlear implants [176] or retinal prostheses [133].

Besides supplying power and transmitting data, inductively coupled links have been noted for their interesting secondary uses. The authors of [116] present a system for joint communication and sensing, in which physical properties of the medium surrounding an RFID tag can be inferred from impedance measurements. For example, the impedance observed at a reader device due to the presence of a tag is influenced by the dielectric properties of the surrounding medium. To further allow sensing of non-dielectric quantities, the authors propose to coat the tag with a sensing film which influences measurements of the tag impedance as a function of the desired medium property. A similar approach is presented in [88], where thin-film sensors impose a variable capacitance or resistance that can be integrated into the resonant antenna circuit of a sensor node.

Two notable examples of secondary applications considered extensively within this thesis are using passive resonant nodes to improve a communication channel, and performing localization, i.e. estimating the unknown position of a node. The state of the art relevant to these topics is discussed in Chapters 4 and 6, respectively.

Highly miniaturized sensor networks using inductive coupling for communication have only rarely been studied in literature, primarily with a focus on technical implementation of the individual nodes. One example is found in [89], where an implantable sensor node integrated on a 8 mm × 3 mm substrate with inductive power supply and communication functionality is demonstrated. The authors of [45] present an integrated node measuring 3.5 mm × 1 mm, in which power supply and downlink communication, i.e. from a central unit to the node, are implemented using an inductively coupled link operating at 866 MHz.

1.3   Contributions and Outline

With its simplicity and the opportunity for synergistic reuse, we identify inductive coupling as a suitable physical layer for large networks of inexpensive, miniaturized nodes. At the same time it is clear that microsensor networks, particularly with a low-complexity physical layer, exhibit limited capabilities in terms of their communication and localization performance. This thesis investigates the applicability of inductive coupling to both macroscale and microscale sensor networks by addressing the question of how some of the limitations inherent to this physical layer can be mitigated. The underlying premise lies in the idea that sensor networks comprised of many nodes can compensate for their limitations by drawing on the resource of node cooperation. We specifically assume that a subset of the nodes present in a network is available to assist a specific task. In the context of wireless communication such nodes are commonly referred to as relays. We study the limits of both communication and localization and investigate the gains that are achievable from the availability of inductively coupled wireless relays.

To the best of our knowledge, both communication and localization have only rarely been studied from the perspective of relay-assisted, inductively coupled ad-hoc sensor networks. Throughout this thesis we will address the following items:

- We investigate how purely passive, inductively coupled relay nodes can help to overcome limitations in communication range and from node misalignment. We show that the observed gains are also achievable for randomly arranged network configurations.

- We show that a network of inductively coupled nodes with limited computational power can employ cooperative communication to implement low-complexity distributed computational abilities. We study the reliability of the resulting computation in the presence of noise and imperfections.

- We demonstrate the possibility of performing accurate localization within inductively coupled networks, even if the nodes to be located are completely passive. To this end we answer questions on the theoretical limits, key parameter values, and achievable localization performance. We also investigate which imperfections must be considered when implementing localization for inductively coupled networks in practice, and how they can be addressed.

- We study the use of supporting nodes to resolve localization ambiguities, and investigate the feasibility of jointly performing communication and localization.

This work is structured in two major parts presenting contributions in the areas of communication and localization, respectively. Before addressing these topics, we conclude the introductory part by providing an overview of linear N-port network theory in Chapter 2 and by discussing the fundamentals of the mechanism of inductive coupling in Chapter 3. These chapters provide the basis for the analysis of communication performance using a circuit-theoretic description of the entire network, which is based on the work in [68] and was modified to account for the unique characteristics of inductively coupled networks. The validity of commonly used assumptions is investigated and scaling laws are derived to understand how the process of miniaturization affects the communication performance in microsensor networks.

Chapters 4 and 5 form the first major part of the thesis which investigates relay assisted communication in inductively coupled microsensor networks. Chapter 4 addresses the limited communication range encountered in inductively coupled networks due to both the rapid decay of the magnetic near field with distance and the influence of misalignment. We investigate passive relaying to overcome this drawback, a technique particularly suitable for low complexity microsensor networks as it employs relay nodes which can be implemented without logic or power supply. We show that gains from passive relaying can also be achieved for random node arrangements—as are typically encountered in sensor networks—if the load impedance of the relay can be chosen adaptively.

Limitations in node complexity, which can be expected for inductively coupled microsensor networks, are the motivation for the wireless artificial neural network scheme proposed in Chapter 5. By reusing the structure of relay assisted communication we show that a sensor network of very low-complexity nodes can compute arbitrary functions of sensor values by implementing an artificial wireless neural network directly on the physical layer. We propose a decentralized gradient-based optimization to choose the required relay gains implementing the artificial neural network, and investigate the viability and robustness of the computation using the basic examples of optical character recognition and the Boolean XOR function.

The second major part of this work in Chapters 6, 7, and 8 covers the topic of reusing inductively coupled communication infrastructure to implement localization in microsensor networks. Chapter 6 introduces the notion of circuit-based near-field localization, which allows for the localization of a completely passive agent node using measurements of circuit parameters such as the antenna input impedance at multiple anchor nodes with fixed and known positions. The localization performance is shown by simulation to adhere well to derived theoretical bounds, and to yield accurate position estimates in the investigated scenarios.

Chapter 7 extends circuit-based localization to systems with only a single anchor performing measurements. This restriction results in ambiguities in the underlying cost function, but is motivated by an envisioned microsensor network architecture in which many low-complexity nodes are controlled by a single central unit with higher complexity. To enable reliable localization in this setting the notion of passive anchors is proposed. We show that these purely passive nodes can resolve the ambiguities either simply by their presence or by switching their load impedances in a synchronized fashion.

Finally, Chapter 8 provides an experimental verification of circuit-based localization. In order to reliably perform localization in a practical setup, the imperfections of the presented measurement systems must be accounted for. We identify the most dominant sources of error to be imperfect knowledge of the involved circuits as well as temporally varying systematic errors of the used measurement instrument. We propose and evaluate efficient calibration schemes for both sources of imperfections and demonstrate that accurate location estimates are also achievable in a practical setup.

2

A Primer in Linear N-port Network Theory

This thesis employs N-port network theory as a model to describe the interaction in inductively coupled communication systems. The goal of this chapter is to provide a summary of the relevant tools that are used throughout the remainder of the work. After introducing the notion of linear N-port networks we discuss port reduction methods, networks with internal sources, noisy networks, and time-variant networks.

2.1 N-port Networks

We consider a general electrical circuit such as the one depicted in Fig. 2.1. The inner structure of this circuit may be unknown and arbitrarily complex, but we assume it is passive, i.e. it contains no sources of energy, and that it consists only of linear circuit elements, i.e. components in which voltage and current follow a linear relation, such as resistors, inductors, or capacitors. The network exhibits a number of *ports*, which are pairs of terminals attached to distinct nodes in the underlying circuit. The ports are identified by a port index $n \in \{1, \ldots, N\}$ and each port is associated with a respective voltage u_n and current i_n which may be static or sinusoid quantities. Energy in form of electrical signals can only enter or leave the network through the ports. The notion of N-port circuit theory allows the abstraction of the underlying circuit, as its external behavior is fully described by the relationship between the voltages and currents at all ports. Due to the assumption of a linear network, the individual voltages and currents must also exhibit a linear relationship, implying that an N-port network can be characterized in the frequency domain by a generally complex $N \times N$ matrix relating all pairwise port quantities at a specific frequency.

There exist several models to characterize a linear passive network. We will mainly use the Z-parameter description. These parameters are defined for each combination of port indices

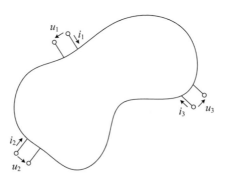

Figure 2.1. Arbitrary N-port network with $N = 3$.

$m, n \in \{1, \ldots, N\}$ as

$$Z_{mn} = \left. \frac{u_m}{i_n} \right|_{i_m = 0} . \tag{2.1}$$

The values Z_{mn} thus physically represent impedances, and are alternatively also referred to as impedance parameters. Arranging all pairwise values in a matrix $\mathbf{Z}$, the N-port is explicitly characterized as

$$\mathbf{u} = \mathbf{Zi} \tag{2.2}$$

where $\mathbf{u} = [u_1, \ldots, u_N]^{\mathrm{T}}$ and $\mathbf{i} = [i_1, \ldots, i_N]^{\mathrm{T}}$, with $(\cdot)^{\mathrm{T}}$ denoting the transpose. It is well known that the Z-parameters are not defined for all networks[1]. As an alternative the Y-parameters—also called admittance parameters—may be employed. They are defined as

$$Y_{mn} = \left. \frac{i_m}{u_n} \right|_{u_m = 0} . \tag{2.3}$$

Written in matrix form, the Y-parameters yield the linear relation

$$\mathbf{i} = \mathbf{Yu}. \tag{2.4}$$

---

[1] A simple example of a network with undefined Z-parameters is a two-port containing a single series resistor.

Left-multiplication of (2.2) by the inverse impedance matrix $\mathbf{Z}^{-1}$ shows that impedance and admittance parameters are related by $\mathbf{Y} = \mathbf{Z}^{-1}$.

For terminated networks operating at RF frequencies, it is often intuitive to describe the port behavior not in terms of voltages and currents, but in terms of the power waves entering or leaving each port. At the nth port these waves are denoted as a_n and b_n, respectively. They are defined as [79]

$$a_n = \frac{u_n + Z_{0,n} i_n}{2\sqrt{\mathfrak{Re}\left\{Z_{0,n}\right\}}}, \text{ and} \tag{2.5}$$

$$b_n = \frac{u_n - Z_{0,n}^* i_n}{2\sqrt{\mathfrak{Re}\left\{Z_{0,n}\right\}}}. \tag{2.6}$$

Here $Z_{0,n}$ represents the characteristic impedance of the line connected to port n of the network[2], and $Z_{0,n}^*$ denotes its complex conjugate. The real part of $Z_{0,n}$ is, by definition, positive. The behavior of the network at port n is then characterized by the reflection coefficient Γ_n, defined as the ratio of a_n and b_n:

$$\Gamma_n = \frac{b_n}{a_n}. \tag{2.7}$$

The reflection coefficient represents a full description of a 1-port network at a given frequency, with the power waves being related to the port current and voltage as

$$u_n = \frac{1}{\sqrt{\mathfrak{Re}\left\{Z_{0,n}\right\}}} \left(Z_{0,n}^* a_n + Z_{0,n} b_n\right), \text{ and} \tag{2.8}$$

$$i_n = \frac{1}{\sqrt{\mathfrak{Re}\left\{Z_{0,n}\right\}}} \left(a_n - b_n\right). \tag{2.9}$$

Extending the notion of (2.7) to an N-port network yields the relation

$$\mathbf{b} = \mathbf{Sa}, \tag{2.10}$$

Here $\mathbf{a} = [a_1, \ldots, a_N]^{\mathrm{T}}$, $\mathbf{b} = [b_1, \ldots, b_N]^{\mathrm{T}}$, and the elements of the matrix $\mathbf{S}$ are referred to as scattering parameters or S-parameters. Intuitively, these values S_{nm} describe the complex amplitude gain of the network from port m to port n. The corresponding power gain of a perfectly matched network is given by $|S_{nm}|^2$. It should be noted that the S-parameters explicitly depend on the specified characteristic impedances of the ports.

Further formulations of the port behavior of linear N-port networks exist, mostly based on

---

[2]The characteristic impedance of transmission lines is often designed to a value of $50\,\Omega$.

the ratios of pairwise port currents and/or voltages. These include hybrid parameters, inverse hybrid parameters, and ABCD parameters. All of these models are formally equivalent such that a network description can be converted between the different representations. Further discussion of these models and their relationships can be found in [117] and [24].

2.2   Port Reduction

A useful circuit-theoretic tool is that of port reduction. By terminating a subset of n of the ports of an N-port network with known, linear loads, the unterminated ports of the network form a linear $(N - n)$-port. To formalize this mechanism we define as partially terminated impedance matrix an impedance matrix $\mathbf{A} \in \mathbb{C}^{N \times N}$ for which ports 1 through $N - n$ are open circuited, while ports $N - n + 1$ through N are terminated by an arbitrary load impedance matrix $\mathbf{B} \in \mathbb{C}^{n \times n}$. The partially terminated structure is depicted in Fig. 2.2,

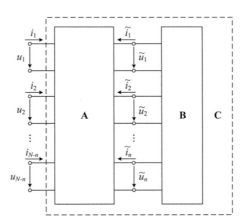

Figure 2.2. Partially terminated N-port network.

with $\mathbf{C} \in \mathbb{C}^{(N-n) \times (N-n)}$ denotes the resulting network at the unterminated ports. Here we assign the vectors $\mathbf{u}$ and $\mathbf{i}$ to denote the voltages and currents at the unloaded ports, while the vectors $\tilde{\mathbf{u}}$ and $\tilde{\mathbf{i}}$ represent the voltages and currents of the terminated ports of $\mathbf{A}$. These quantities must fulfill the condition

$$\begin{bmatrix} \mathbf{u} \\ \tilde{\mathbf{u}} \end{bmatrix} = \begin{bmatrix} \mathbf{A}_{11} & \mathbf{A}_{12} \\ \mathbf{A}_{21} & \mathbf{A}_{22} \end{bmatrix} \begin{bmatrix} \mathbf{i} \\ \tilde{\mathbf{i}} \end{bmatrix}, \tag{2.11}$$

where $\mathbf{A}_{11}$, $\mathbf{A}_{12}$, $\mathbf{A}_{21}$, and $\mathbf{A}_{22}$ are the constituent submatrices of $\mathbf{A}$. To find the relationship between the voltages and currents at the unloaded ports, we write the relation in (2.11) explicitly as

$$\mathbf{u} = \mathbf{A}_{11}\mathbf{i} + \mathbf{A}_{12}\tilde{\mathbf{i}}, \quad \text{and} \tag{2.12}$$
$$\tilde{\mathbf{u}} = \mathbf{A}_{21}\mathbf{i} + \mathbf{A}_{22}\tilde{\mathbf{i}}. \tag{2.13}$$

At the same time the partial termination by the load matrix $\mathbf{B}$ imposes the constraint

$$\tilde{\mathbf{u}} = -\mathbf{B}\tilde{\mathbf{i}}. \tag{2.14}$$

By equating (2.14) and (2.13) we obtain

$$\tilde{\mathbf{i}} = -(\mathbf{A}_{22} + \mathbf{B})^{-1}\mathbf{A}_{21}\mathbf{i}, \tag{2.15}$$

which can be substituted back into (2.12) to find a general expression of the matrix $\mathbf{C}$ relating the unloaded voltages and currents as

$$\mathbf{C} = \mathbf{A}_{11} - \mathbf{A}_{12}(\mathbf{A}_{22} + \mathbf{B})^{-1}\mathbf{A}_{21}. \tag{2.16}$$

The matrix $\mathbf{C}$ is generally referred to as input impedance matrix. By interchanging voltages and currents in the calculations above it can be seen that (2.16) also holds for partially terminated admittance matrices.

An important special case of the input impedance matrix is given for two-port networks with an intended direction of signal flow, such as a filter network with an input and output port. Without loss of generality we define port 1 as input and port 2 as output. In this context the input impedance Z_{in} of the two-port with impedance matrix $\mathbf{Z}$ is defined as

$$Z_{\text{in}} = Z_{11} - \frac{Z_{12}Z_{21}}{Z_{22} + Z_{\text{L}}}. \tag{2.17}$$

It describes the impedance observed at an open-circuited input when the output is terminated with a load Z_{L}. Likewise, if the input port is terminated by a source impedance Z_{S}, the output impedance is defined as

$$Z_{\text{out}} = Z_{22} - \frac{Z_{12}Z_{21}}{Z_{11} + Z_{\text{S}}}. \tag{2.18}$$

The notion of the input impedance of a port n can be related to the reflection coefficient as defined in (2.7) by

$$\Gamma_n = \frac{b_n}{a_n} = \frac{Z_{\text{in},n} - Z_{0,n}}{Z_{\text{in},n} + Z_{0,n}}, \tag{2.19}$$

signifying that the ratio of reflected and impinging power waves is a measure of the mismatch between the input impedance of the port and its corresponding characteristic line impedance. By implication a port can be matched to exhibit no reflections, which is the case for $Z_{\text{in},n} = Z_{0,n}$.

2.2.1 Port Reduction for S-Parameter Measurement

In the scattering parameter domain, the concept of port reduction is implicitly included in the definition of the network parameters, as it is assumed that the N-port network of interest is terminated with the characteristic impedances of the respective ports. To illustrate this point, we consider the measurement of the parameter S_{mn} of an N-port network, which can be calculated from a response b_m at port m to an excitation a_n at port n. This measurement can be obtained using a two-port vector network analyzer (VNA) connected to the respective ports. If the remaining $N-2$ ports have arbitrary terminations, the response b_m is found as

$$b_m = \sum_{i=1}^{N} S_{mi}a_i. \tag{2.20}$$

Here, the waves a_i entering the N-port at ports $i \neq n, m$ are generated by reflections from the loads at the respective terminated ports. These loads can be interpreted as one-port networks having the reflection coefficients

$$\Gamma_{\text{L},i} = \frac{a_i}{b_i} \tag{2.21}$$

with a_i and b_i being defined as the waves entering and leaving port i of the N-port network. By ideally matching the terminations at all ports of the network, it follows from (2.19) that the reflection coefficients of the terminations become $\Gamma_{\text{L},i} = 0, i \in 1, \ldots, N$, and accordingly

$a_i = 0, i \neq n$. The relation in (2.20) is then reduced to

$$b_m = S_{mn}a_n.$$
(2.22)

Following this reasoning, an N-port network described by $\mathbf{S}$ and ideally matched at $N - 2$ ports is inherently reduced to the two-port described by the submatrix

$$\mathbf{S}_{mn} = \begin{bmatrix} S_{mm} & S_{mn} \\ S_{nm} & S_{nn} \end{bmatrix}.$$
(2.23)

Measuring these submatrices for all port combinations (m, n) allows for the full measurement of $\mathbf{S}$ using a two-port VNA. However, perfect matching is difficult to achieve in practice [90]. The characterization of N-port networks with imperfectly matched partial terminations is beyond the scope of this work, but has been investigated extensively in literature (cf. [160], [90], [122] and the references therein).

An important class of linear N-port networks is given by reciprocal networks which have the property $S_{mn} = S_{nm}$. For these networks the submatrix $\mathbf{S}_{mn}$ can be obtained from a sequence of measurements at one of the ports of interest, while using different terminations at the other port. These terminations have an associated reflection coefficient $\Gamma_{\mathrm{L},n} = a_n/b_n$. We obtain the response b_m from an excitation at port m as

$$\begin{aligned} b_m &= S_{mm}a_m + S_{mn}a_n \\ &= S_{mm}a_m + S_{mn}\Gamma_{\mathrm{L},n}b_n. \\ &= \left(S_{mm} + \frac{S_{mn}\Gamma_{\mathrm{L},n}S_{nm}}{1 - S_{nn}\Gamma_{\mathrm{L},n}} \right) a_m. \end{aligned}$$
(2.24)

It follows that the value measured by the VNA at port m is given by

$$\begin{aligned} \frac{b_m}{a_m} &= S_{mm} + \frac{S_{mn}S_{nm}\Gamma_{\mathrm{L},n}}{1 - S_{nn}\Gamma_{\mathrm{L},n}} \\ &= S_{mm} + \frac{b_m}{a_m}S_{nn}\Gamma_{\mathrm{L},n} - \Gamma_{\mathrm{L},n} \cdot (S_{mn}S_{nm} - S_{mm}S_{nn}). \end{aligned}$$
(2.25)

It can be seen that (2.25) forms a linear system of equations with the unknowns S_{mm}, S_{nn}, and $(S_{mn}S_{nm} - S_{mm}S_{nm})$, the solution of which fully determines the reciprocal two-port network assuming three independent values of $\Gamma_{\mathrm{L},n}$ are given.

23

2.3 Networks with Internal Sources

We next consider a linear one-port network containing an arbitrary number of energy sources in the form of voltage or current sources. Any excitation within such an *active* network will lead to an open-circuit voltage at the unterminated port. The linearity of the network allows to find the overall port voltage as the superposition of the port voltages which result from the excitations of all individual sources within the network.

If the port is terminated with a load such that a current is allowed to flow, the port voltage and current are related by the input impedance Z_{in} of the port. From the perspective of the terminating load, the one-port network therefore behaves equivalently to a voltage source. This property of one-port networks is described by Thévenin's theorem [24], which states that the port behavior of *any* linear one-port network containing voltage and/or current sources can be replaced by an equivalent voltage source in series with an internal impedance[3].

Thévenin's theorem allows simplified network analysis, as can be seen from an exemplary communication system depicted on the left-hand side in Fig. 2.3. The communication system generates a signal using a voltage source with amplitude u_S and internal resistance R_S. The

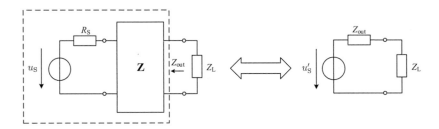

Figure 2.3. Application of Thévenin's theorem for exemplary communication system. The source and two-port network can be summarized, as indicated by the dashed box, and replaced by an equivalent source.

signal is passed through a two-port network (e.g. representing a filter) to a receiver modeled by a load impedance Z_L. In this setting the two-port network—described at each specific frequency by an impedance parameter matrix $\mathbf{Z}$—and the source can be replaced by their Thévenin equivalent. The voltage u'_S of the equivalent source is given by the open-circuit

---

[3]Following a similar reasoning, Norton's theorem states that any one-port with internal sources is equivalent to a current source with a parallel impedance. It is not used in this thesis.

voltage at the output of the two-port resulting from the excitation of the internal source. If the output port is unterminated, we find

$$u_S = i_1 \left(R_S + Z_{11} \right).$$ (2.26)

Using the condition that $i_2 = 0$, we find the equivalent source voltage as

$$u_S' = \frac{Z_{12}}{R_S + Z_{11}} \cdot u_S.$$ (2.27)

The internal impedance of the equivalent source is given by the output impedance of the two-port terminated by the internal source at its input port. It is calculated as given in (2.18). By using the equivalent source, the communication system can be described using the simple model depicted on the right-hand side in Fig. 2.3.

Thévenin's theorem is formulated for active linear one-port networks, but may be generalized to networks with internal sources having any number of ports. To this end, an active N-port network is equivalently represented by a passive network[4] with external voltage sources on each port having amplitudes identical to the open-circuit voltages resulting from internal excitation.

2.4 Noisy Networks

Noise plays an important role in the analysis of electrical circuits, particularly in a communication context where noise imposes fundamental limitations on communication performance [161]. We define electrical noise as voltages or currents inherent to the network which exhibit an apparently random temporal behavior which effectively cannot be predicted due to complex, unknown, or truly random sources. There exist several noise producing mechanisms within linear electrical networks. The most important noise types are *thermal noise*, which is the result of the thermal excitation of the electrons in the circuit, *shot noise*, which consists of pulses of current stemming from the quantized flow of electrons, e.g. from passing through a potential barrier in a semiconductor junction, and *1/f noise*, the sources of which are not fully understood [60]. A detailed investigation of these noise mechanisms is outside of the scope of this thesis; for further discussion the reader is referred to [92] and the references therein.

---

[4]The passive network is found by short-circuiting all internal voltage sources and open-circuiting all internal current sources.

Particularly in the analysis of circuits describing communication systems, deterministic signals that are unrelated to the signal of interest—e.g. interference from secondary sources—are sometimes also interpreted as noise. In a circuit-theoretic context noise can be modeled by voltage or current sources with amplitudes given by a random process. In this sense the previous considerations for networks with deterministic internal sources also hold for noisy networks, with the distinction that the complex amplitudes of the noise signals cannot be superimposed directly due to their randomness. To analyze communication in the presence of noise sources, it is therefore necessary to characterize them in a statistical sense, e.g. by their distribution, correlation, noise power spectral density (PSD), or noise root mean square (RMS) voltage.

In this thesis the most important noise mechanism is thermal noise, which can be modeled by a Gaussian distribution with zero mean and variance σ_N^2. It has been shown in [108] that the spectral density of the available thermal noise power generated in a conductor is given by

$$N_0 = k_B T \left(\frac{\frac{h_P f}{k_B T}}{\exp\left(\frac{h_P f}{k_B T}\right) - 1} \right). \tag{2.28}$$

where k_B is the Boltzmann constant, T is the temperature of the conductor, h_P is Plank's constant, and f is the frequency of interest. For the temperatures and frequencies of interest in wireless communications the condition $h_P f \ll k_B T$ holds[5], and (2.28) can be simplified to the well-known form

$$N_0 \approx k_B T. \tag{2.29}$$

Here $T = 290\,\mathrm{K}$ is used in literature as standard noise temperature, resulting in a thermal noise PSD of $-174\,\mathrm{dBm/Hz}$. Following Thévenin's theorem, the noisy conductor can be represented as equivalent one-port having a noise voltage source with random amplitude u_N and internal resistance identical to the resistance R of the conductor. Intuitively, the noisy resistive conductor is therefore modeled as noiseless resistor in series with a noise source. By terminating the noise source using a matched resistance to extract the maximum power, we can equate the available power in a bandwidth Δf to the expression

$$k_B T \Delta f = \frac{\mathsf{E}\left[|u_N^2|\right]}{4R}. \tag{2.30}$$

The thermal noise power of a conducting, resistive element in a bandwidth Δf is therefore

---

[5]At room temperature, the condition $h_P f \ll k_B T$ is valid for frequencies up to approximately $600\,\mathrm{GHz}$ [37].

given as

$$\sigma_N^2 = \mathsf{E}\left[|u_N^2|\right] = 4k_BT\Delta f R. \tag{2.31}$$

Similar to N-ports with internal sources Thévenin's theorem again generalizes to noisy N-ports, allowing to replace them with a noiseless N-port and an equivalent noise voltage sources at each port. We write these equivalent noise voltages as vector $\mathbf{u}_N = [u_{N1}, \dots, u_{NN}]^T$. It was shown in [163] that the covariance matrix $\mathbf{\Sigma}_N = \mathsf{E}\left[\mathbf{u}_N\mathbf{u}_N^H\right]$ of a noisy N-port with impedance parameter matrix $\mathbf{Z}$ is given by

$$\mathbf{\Sigma}_N = 4k_BT\Delta f\mathfrak{Re}\left\{\mathbf{Z}\right\}. \tag{2.32}$$

2.5 Time-Variant Networks

Until now we have assumed that the considered N-port networks are time-invariant. However, many practical communication systems contain time-variant elements. For these types of networks, the presented frequency-domain description of the circuits does not hold in general [15]. In these cases, an analysis of the network can be obtained from a set of differential equations describing the underlying circuit. A set of specialized theories has been developed for the description of time-variant networks for which a characterization by differential equations is not possible (e.g. when the variations are random in nature) or too tedious [177], [28].

A special case of time-variant networks relevant for inductively coupled communication systems lies in networks which contain a time-variant binary switching. Switching is the basis of the load modulation scheme used in RFID systems and described in Section 3.2. In this scheme a communication signal is generated in a wireless node by the switching of a load impedance according to a desired bit pattern. A simple way to analyze such networks is to treat them as block-wise time-invariant network, for which any transients resulting from the switching are ignored. However, duration of the transients is on the order of the switching intervals, they must be taken into account, e.g. from analysis of differential equations. Networks of switched capacitors have received particular attention in literature due to their ability to provide an accurate and tunable emulation of continuous-time filters in the discrete-time domain [7]. A number of methods have been developed for the analysis of this type of networks (cf. [162] and the references therein). As an example, for two-port networks containing periodically switched capacitors the authors of [81] present a z-domain analysis method based on the charge-storage properties of the capacitors. As a result, it is shown that the time-variant two-port can equivalently be represented by a time-invariant four-port network with input and output ports for even and odd switch states, respectively, and that the resulting network can be further analyzed using linear network theory. This approach is generalized to N-port networks in [59].

3

Inductive Coupling as Physical Layer

In this chapter, a systematic description of inductively coupled sensor networks is provided. We first review the mechanism of inductive coupling. Using multiport circuit theory, we subsequently develop a formal model of both signal and noise behavior in the network which forms the basis for analysis in the subsequent chapters.

3.1 Inductive Coupling

Our interest lies in the mechanism of inductive coupling, which presents a means for interaction between a pair of wireless nodes. As all electromagnetic phenomena, a formal description of inductive coupling can be derived from Maxwell's equations, which are given as follows.

$$\nabla \times \mathbf{E} = -\frac{\partial \mathbf{B}}{\partial t} \tag{3.1}$$

$$\nabla \times \mathbf{B} = \mu \left(\varepsilon \frac{\partial \mathbf{E}}{\partial t} + \mathbf{J} \right) \tag{3.2}$$

$$\nabla \cdot \mathbf{E} = \frac{\rho_Q}{\varepsilon} \tag{3.3}$$

$$\nabla \cdot \mathbf{B} = 0 \tag{3.4}$$

Herein $\mathbf{E}(\mathbf{r}, t, \omega)$ and $\mathbf{B}(\mathbf{r}, t, \omega)$ denote the electric field strength and magnetic flux density, respectively. Furthermore, $\mathbf{J}(\mathbf{r}, t, \omega)$ is the current density and $\rho_Q(\mathbf{r}, t, \omega)$ the charge density. While all quantities are dependent on spatial position $\mathbf{r}$, time t, and frequency $\omega = 2\pi f$, these dependencies will be omitted in the following for notational convenience. Throughout this thesis, we will assume a linear, isotropic and homogeneous medium with permittivity ε and permeability μ.

The principle of induction is governed by Faraday's law in (3.1). From its integral form it can be intuitively seen that a temporally changing magnetic flux density $\mathbf{B}$ induces a voltage u_{ind} along any curve denoted by $\delta\Sigma$, and that the magnitude of the voltage is given by the rate of change of the total flux through any surface Σ spanned by the curve:

$$u_{\text{ind}} = \oint_{\delta\Sigma} \mathbf{E} \cdot \mathrm{dl} = -\frac{\partial}{\partial t} \iint_{\Sigma} \mathbf{B} \mathrm{da}. \tag{3.5}$$

Two closely spaced conductors experience inductive coupling on the basis of this principle, as a changing current in one conductor generates a time variant magnetic field according to Ampère's law in (3.2), which in turn induces a voltage in the other conductor. This mechanism allows for the transmission of both power and information from one wireless node to another.

The design of inductively coupled networks requires a quantitative analysis of the electromagnetic interaction. Analytically finding solutions to Maxwell's equations for practical systems is often not feasible except for very simple cases. Instead, approximations of the analytical solutions can be found by numerical methods, which employ a spatial discretization of the entire or parts of the configuration of interest. Numerical techniques for electromagnetic modeling found in literature include the finite element method (FEM), the method of moments (MOM), or the finite difference time domain method (FDTD). An overview of these numerical methods in the context of electromagnetic analysis is provided in [63] and [19].

As an alternative to numerical analysis, analytical solutions can often feasibly be found when adequate simplifications are used. In the following, we provide a model of the inductive coupling between a pair of node antennas. To this end we make use of two approximations. On one hand, we can assume the conductors of the node antennas are thin enough to be considered as infinitely thin. This implies that the current distribution in the antennas is reduced to a one-dimensional quantity. As such, the spatial integration over the field sources is simplified. Additionally we employ a magnetoquasistatic (MQS) approximation of Maxwell's equations. In this regime all field-producing source currents can approximately be treated as time-invariant direct current (DC) sources. Specifically, for sufficiently slowly changing electric fields (i.e. at low frequencies) the displacement current in (3.2) can be neglected, i.e. $\varepsilon\frac{\partial \mathbf{E}}{\partial t} + \mathbf{J} \approx \mathbf{J}$, implying that radiation does not occur. The MQS approximation allows to derive simplified expressions for the quantities of self and mutual inductance which are the basis of the system models considered in this work. It is justified if the frequencies of interest are low enough for the corresponding wavelengths to be considerably greater than the physical dimensions of the setup, as discussed in Section 3.3.

3.1.1 Definition of Coordinate System

Throughout this thesis, we assume to have a sensor network consisting of N nodes which in the most general form may be arranged with arbitrary positions and orientations. The position of the nth node in the network is given by the vector $\mathbf{p}_n = [x_n, y_n, z_n]^\mathrm{T}$, where x_n, y_n, and z_n represent the respective node position in Cartesian coordinates. Likewise, the orientation is defined by the vector $\mathbf{q}_n = [\alpha_n, \beta_n, \gamma_n]^\mathrm{T}$ which contains three angles denoting rotation around the three axes of a local coordinate system located at the respective node's position. In order to associate the node position $\mathbf{p}_n$ to a point of the physical wireless device, all node antennas are assumed to be flat and of circular shape. With this assumption we may define that the node position corresponds to the center point of the antenna, as shown in Fig. 3.1.

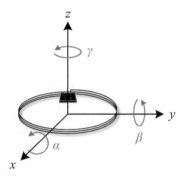

Figure 3.1. Local coordinate system of wireless node with definition of rotation axes. The origin of the local coordinate system is centered on the position $\mathbf{p}_n$ of the node.

It may sometimes be convenient to define the positions of the nodes in spherical coordinates, e.g. for the description of electromagnetic phenomena. In this case we may use $\mathbf{p}_n = [\rho_n, \vartheta_n, \varphi_n]^\mathrm{T}$, where the vector components denote radius, inclination, and azimuth, respectively. We summarize the degrees of freedom (DOF) for the arrangement of each node in a parameter vector $\boldsymbol{\theta}_n = \left[\mathbf{p}_n^\mathrm{T}, \mathbf{q}_n^\mathrm{T}\right]^\mathrm{T}$. In the most general case the vectors $\boldsymbol{\theta}_n$ take on values in $\mathbb{R}^6$ corresponding to the 6DOF parameter space of each node's placement. It should be noted that when assuming circular symmetry of the antenna around the z-axis, the parameter space of each node is reduced to five degrees of freedom.

3.1.2 Self- and Mutual Inductance in the MQS Regime

We begin by considering a single node equipped with a loop antenna. The conductor of this antenna encloses the surface Σ. According to Ampère's law (3.2), a current in the conductor of the antenna will cause the magnetic flux

$$\Psi(i) = \int_\Sigma \mathbf{B}(i)\mathrm{d}\mathbf{a} \tag{3.6}$$

to flow through the surface Σ. For the description of inductive coupling, it is convenient to define the self-inductance L of the antenna as

$$L = \frac{\Psi(i)}{i}. \tag{3.7}$$

If the antenna loop is made up of multiple windings of the conductor, we assume that its ν windings are spaced sufficiently close together such that the total magnetic flux is given by the sum of flux contributions $\Phi(i)$ from a single winding as $\Psi(i) = \nu\Phi(i)$.

The definition of self-inductance in (3.7) may be extended to multiple antennas to obtain the notion of mutual inductance. For a pair of antennas denoted by the indices m and n the mutual inductance M_{mn} is defined as

$$M_{mn} = \frac{\Psi_n(i_m)}{i_m}. \tag{3.8}$$

In this context, antenna m is often referred to as primary antenna, and antenna n as secondary antenna.

Using the notions of self- and mutual inductance, we can quantify the relationship between voltages and currents in a pair of coupled antennas by modeling them as linear two-port. A visualization of the underlying circuit formed by an antenna pair is given in Fig. 3.2. This model is equivalent to that of an ideal air-cored transformer. In general, both time-varying currents i_m and i_n will induce voltages in both antennas according to Faraday's law of induction in (3.5). When calculating the induced voltage for the coupled antennas, the

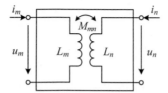

Figure 3.2. Two-port model for inductively coupled lossless antennas.

magnetic flux density $\mathbf{B}$ and its corresponding total magnetic flux Ψ are the superposition of the fluxes generated by the individual currents i_m and i_n. To obtain an expression for the port voltages u_m and u_n, we employ the definitions (3.7) and (3.8) as well as the notion of harmonic currents. For the example of antenna m, the port voltage is calculated as

$$u_m = -\frac{\partial \Psi}{\partial t} = -j\omega\Psi = j\omega L_m i_m + j\omega M_{mn} i_n. \tag{3.9}$$

Following the same approach for antenna n, the relationship of port currents and voltages of the coupled antennas can therefore be represented in the impedance parameter description $\mathbf{u} = \mathbf{Z}\mathbf{i}$, as discussed in Chapter 2. The two-port relation takes the explicit form

$$\begin{bmatrix} u_m \\ u_n \end{bmatrix} = j\omega \begin{bmatrix} L_m & M \\ M & L_n \end{bmatrix} \cdot \begin{bmatrix} i_m \\ i_n \end{bmatrix}. \tag{3.10}$$

This description can be extended to arbitrary number of antennas. For a system of N coupled antennas, each port voltage will contain a self-inductance component and $N-1$ contributions from mutual inductances:

$$\begin{bmatrix} u_1 \\ \vdots \\ u_N \end{bmatrix} = j\omega \begin{bmatrix} L_1 & \cdots & M_{1N} \\ \vdots & \ddots & \vdots \\ M_{N1} & \cdots & L_N \end{bmatrix} \cdot \begin{bmatrix} i_1 \\ \vdots \\ i_N \end{bmatrix}. \tag{3.11}$$

The values of the mutual inductance in (3.11) can be calculated for each pair of antennas by inserting (3.6) into (3.8), where the magnetic flux from sources other than the antenna of interest is disregarded in the calculation. Taking into account the multiple windings of both antennas, we obtain the mutual inductance from the superposition of the contributions of the individual windings as

$$M_{mn} = \frac{\nu_m \nu_n}{i_m} \int_{\Sigma_n} \mathbf{B}(i_m) \mathrm{da}, \qquad (3.12)$$

where $\mathbf{B}(i_m)$ here represents the magnetic flux density produced by a single winding of antenna m.

By employing the notion of the magnetic vector potential $\mathbf{A}$ with the defined property $\mathbf{B} = \nabla \times \mathbf{A}$, the integral in (3.12) can be expanded as

$$\int_{\Sigma} \mathbf{B} \cdot \mathrm{da} = \int_{\Sigma} (\nabla \times \mathbf{A}) \cdot \mathrm{da} = \oint_{\delta\Sigma} \mathbf{A} \cdot \mathrm{dl}, \qquad (3.13)$$

where we have used Stokes' theorem to obtain the line integral along the surface boundary $\delta\Sigma$. In general, the magnetic vector potential $\mathbf{A}(\mathbf{r})$ resulting from an arbitrary current density $\mathbf{J}(\mathbf{r})$ is given by

$$\mathbf{A}(\mathbf{r}) = \frac{\mu}{4\pi} \int_{\Omega} \frac{\mathbf{J}(\mathbf{r}')}{\|\mathbf{r} - \mathbf{r}'\|} \mathrm{dr}', \qquad (3.14)$$

where the integration is performed over the volume Ω enclosing the source current. For the specific case of a thin-wire loop antenna, the current density $\mathbf{J}(\mathbf{r}')$ is approximately one-dimensional. With this consideration and by combining (3.12) and (3.14), we obtain the well-known double integral expression for the mutual inductance of thin-wire inductors[1], which is given as

$$M_{mn} = \frac{\mu}{4\pi} \oint_{\delta\Sigma_m} \oint_{\delta\Sigma_n} \frac{\mathrm{dl}_m \cdot \mathrm{dl}_n}{\|\mathbf{l}_m - \mathbf{l}_n\|}. \qquad (3.15)$$

It is worth noting that the mutual inductance is therefore solely dependent on the geometries and relative positioning of the curves $\delta\Sigma_m$ and $\delta\Sigma_n$ describing the antenna loops. Furthermore it can be seen that the order of integration can be interchanged, implying the reciprocity of mutual inductance: $M_{mn} = M_{nm}$.

The mutual inductance is related to the respective self-inductances by means of the coupling coefficient $k_{mn} \in [-1, 1]$ as

$$M_{mn} = k_{mn} \sqrt{L_m L_n}. \qquad (3.16)$$

---

[1] The expression (3.15) is often referred to as Neumann formula on the basis of the work in [107].

The coupling coefficient can be interpreted as the ratio of $\Psi_n(i_m)$, the magnetic flux penetrating the secondary antenna, to $\Psi_m(i_m)$, the magnetic flux generated by the primary antenna. The similarity of the definitions of self- and mutual inductance in (3.7) and (3.8) suggests that (3.15) can also be used to calculate the self-inductance of an antenna by setting $n = m$. However, in this case the wire filaments coincide, leading to an undefined integral as $\|\mathbf{l}_m - \mathbf{l}_n\|$ vanishes. It is shown in [29] that by taking a nonzero wire radius a into account, the self inductance can be calculated by evaluating (3.15) only at non-overlapping points, and evaluating the remaining contributions explicitly. For inductors of length $l \gg a$ the self inductance then results to

$$L = \left(\frac{\mu}{4\pi} \oint_{\delta\Sigma} \oint_{\delta\Sigma'} \frac{\mathbf{dr} \cdot \mathbf{dr'}}{\|\mathbf{r} - \mathbf{r'}\|} \right)_{\|\mathbf{r}-\mathbf{r'}\|>a/2} + \frac{\mu}{4\pi} l \Upsilon + \dots, \tag{3.17}$$

where negligible terms for $l \gg a$ have been omitted and the parameter Υ depends on the distribution of the current in the cross section of the conductor. If the current can be assumed to be uniformly distributed along the cross section $\Upsilon = 0.5$, if the current flows only in the surface of the wire then $\Upsilon = 0$.

The direct integration of (3.15) is often infeasible except for special cases with high degrees of symmetry. However, exact and approximate solutions have been found for many canonical thin-wire geometries in literature. A large collection of cases for circular loop antennas and various other regular inductor geometries is provided in [10, 11, 48, 49, 149, 172] and the references therein.

3.1.3 Dipole Approximation of Mutual Inductance

A particularly simple description of the mutual inductance of two loop antennas can be derived if the dimensions of the involved antenna loops are small compared to the distance d between them. To derive this approximation we first consider the magnetic field generated by a single thin-wire loop antenna with radius r_m and number of windings ν_m. Without loss of generality we place this antenna at the origin. A harmonic current with amplitude i_m flows through the antenna conductor, such that a magnetic field $\mathbf{B}(\mathbf{r})$ is generated according to Ampère's law at each spatial point $\mathbf{r}$. It can be shown [125] that for points $\mathbf{r}$ sufficiently far away from the antenna position, i.e. $\|\mathbf{r}\| \gg r_m$, the magnetic field becomes independent of the size of the loop and can be calculated in spherical coordinates as

$$
\mathbf{B} = \begin{bmatrix} B_\rho \\ B_\vartheta \\ B_\varphi \end{bmatrix} = \|\mathbf{m_d}\| \frac{\mu}{4\pi\rho^3} \begin{bmatrix} 2\cos\vartheta \\ \sin\vartheta \\ 0 \end{bmatrix}, \tag{3.18}
$$

where the vector $\mathbf{m_d} = i_m\nu_m\mathbf{a}_m$, called the magnetic dipole moment, is the product of the antenna's current, windings, and the surface area vector spanned by the conductor loop.

The expression of the magnetic flux density can be used to calculate the mutual inductance with a second antenna at position $\mathbf{r}$. The mutual inductance of the antenna pair can be calculated according to (3.12) for which we evaluate the flux $\Psi_n(i_m)$ through the second antenna. For distant antenna positions such that $\|\mathbf{r}\| \gg r_n$, we can assume that the magnitude of the magnetic field flowing through antenna n is nearly constant over it's surface $\mathbf{a}_n$. We can therefore find the approximate the flux as

$$
\Psi_n(i_m) \approx \mathbf{B}(\mathbf{r}) \cdot \nu_n\mathbf{a}_n. \tag{3.19}
$$

The resulting approximation of the mutual inductance is given as

$$
M_{mn} = \mu\pi\nu_m\nu_n \frac{r_m^2 r_n^2}{4d_{mn}^3} \cdot J_{mn}, \tag{3.20}
$$

where d_{mn} is the distance between the antennas and J_{mn} is a polarization factor which depends on the relative orientation of the antennas. The latter is given as in [75]

$$
J_{mn} = 2\sin(\theta_m)\sin(\theta_n) + \cos(\theta_m)\cos(\theta_n)\cos(\phi) \tag{3.21}
$$

Here, θ_m and θ_n are the angles between the respective surface vectors—$\mathbf{a}_m$ and $\mathbf{a}_n$—and $\mathbf{r}$, the vector connecting the antenna positions. Furthermore ϕ is calculated by projecting $\mathbf{a}_m$ and $\mathbf{a}_n$ to the plane with surface normal $\mathbf{r}/\|\mathbf{r}\|$ and calculating the angle between the resulting vectors.

The magnitude of the polarization factor is bounded to $|J_{mn}| \leq 2$. This bounding value, which accordingly also maximizes the magnitude of the mutual inductance, is found by first maximizing (3.21) with respect to ϕ, and noting that the resulting summands are orthogonal both in θ_m and θ_n, with the joint maximum being $J_{mn} = 2$ achieved for $\theta_m = \theta_n = \pi/2$. Subsequent minimization of the polarization factor following the same procedure yields $J_{mn} = -2$ for $\theta_m = \pm\pi/2$ and $\theta_n = -\theta_m$. Both the minimum and the maximum value of J_{mn} therefore correspond to cases where both antennas are coaxially arranged. The values of the extrema are independent of ϕ which, in fact, has no meaningful value for coaxial arrangements.

3.1.4 Comparison of Inductance Models

The dipole approximation in (3.20) provides a simple and efficient model for the calculation of the mutual inductance. However, its underlying assumption of both antennas being smaller than their mutual distance breaks down for closely spaced configurations, leading to a mismatch between the analytical expression in (3.15) and the dipole model of the mutual inductance. Furthermore it should be noted that the analytical solution is also an approximation as it uses the assumption of an infinitely thin conductor.

It is therefore desirable to gauge the validity of both expressions. To obtain suitable reference values, we use the inductance simulation program FastHenry [72]. This tool can be used to obtain the mutual inductance between arbitrarily shaped 3D conductor arrangements by spatial discretization of Maxwell's equations in the MQS regime, which are then efficiently solved using mesh analysis and multipole acceleration [47].

To compare the inductance models we consider a setup of two coaxially arranged loop antennas with a radius of $r_m = r_n = 2.5\,\text{cm}$ and $\nu_m = \nu_n = 1$ winding, as indicated in Fig. 3.3. We are interested calculating the mutual inductance as the distance d between the antennas is varied.

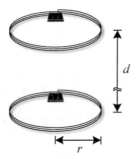

Figure 3.3. Coaxial setup for comparison of inductance models.

The reference values for the mutual inductance are calculated with FastHenry using a 3D antenna model having a square wire cross section with a side length of $150\,\mu\text{m}$. The results are shown in Fig. 3.4 for distances from $0.2\,\text{cm}$ to $40\,\text{cm}$. Here the reference solution corresponds to the dashed red curve.

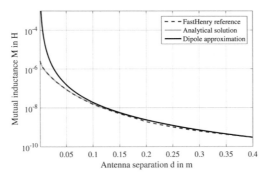

Figure 3.4. Numerical evaluation of mutual inductance for coaxial arrangement of two loop antennas.

For the chosen antenna geometry with a thin-wire assumption, the Neumann integral in (3.15) has the analytical solution [119]

$$M_{mn} = \mu \nu_m \nu_n \sqrt{r_m r_n} \left\{ \left(\frac{2}{\kappa} - \kappa \right) K(\kappa) - \frac{2}{\kappa} E(\kappa) \right\}, \qquad (3.22)$$

where $K(\kappa)$ and $E(\kappa)$ are the complete elliptic integrals of the first and second kind, respectively. Their parameter κ, not to be confused with the coupling coefficient k, is calculated as

$$\kappa = \sqrt{\frac{4 r_m r_n}{d^2 + (r_n + r_m)^2}}. \qquad (3.23)$$

This result corresponds to the green line in Fig. 3.4 and matches very well with the reference, indicating that the thin-wire approximation introduces no significant error for the chosen, practical dimensions.

Finally, the results obtained from the dipole approximation are shown in black. As can be expected they generally show a good match for large antenna separations, in this case larger than $10\,\mathrm{cm}$. The relative error of the dipole approximation with respect to the reference solution is $25.5\,\%$ at $d = 10\,\mathrm{cm}$, $74.1\,\%$ at $d = 5\,\mathrm{cm}$, and $206.5\,\%$ at $d = 3\,\mathrm{cm}$. At closer distances, the results obtained from the dipole approximation quickly become unphysical as (3.20) approaches infinity for $d \to 0$. We conclude from these results that the dipole approximation provides a simple and sufficiently accurate representation of the mutual inductance for all but very close pairwise arrangements and can therefore be used as a computationally efficient coupling model, which is particularly beneficial in the analysis of inductively coupled networks containing a large number of nodes. However, when using the dipole approximation the overemphasis of the coupling for closely spaced arrangements must be taken into account.

3.1.5 Resistive Loss Mechanisms

Until now we have implicitly assumed ideal lossless antennas. In this context we define antenna losses as the undesired dissipation of energy by the antenna. In the description of antennas one generally differentiates two loss mechanisms, namely thermal losses and radiative losses. Thermal losses encompass the thermal dissipation of power due to conductive and dielectric properties of the antenna. Radiation losses, on the other hand, model the power dissipation in the form of electromagnetic radiation, which is undesired as inductive coupling is an effect of the reactive near-field. Both loss mechanisms can be modeled in a circuit-theoretic antenna description by resistive components [12]. To this end we model thermal losses by the loss resistance R_Ω, and radiative power losses by the radiation resistance R_R.

To quantify the value of the losses, we consider a thin antenna of turn radius r, wire radius a, and wire conductivity σ. The loss resistance R_Ω depends on the current distribution within the antenna wire, which in turn is a function of the frequency ω. At DC and quasi-DC frequencies, the current density in a thin conductor can be assumed to be uniformly distributed along its cross section [91]. The DC loss resistance of a single-turn (i.e. $\nu = 1$) loop antenna can therefore be given as

$$R_\Omega|_{\omega=0} = \frac{2r}{a^2\sigma}. \tag{3.24}$$

For higher frequencies the current density becomes increasingly concentrated at the boundary layer between the conductor and its surrounding dielectric due to eddy currents inside the conductor, a behavior well known in literature as skin effect [55]. The resulting current distribution, which decreases exponentially with depth within the conductor, is often approximated as the current being uniformly distributed on a hollow cylinder with outer radius r and a thickness δ called the skin depth, which is defined as the depth at which the amplitude of the current density vector has decreased to $1/e$ of the value at the surface. The skin depth is calculated as [117]

$$\delta = \sqrt{\frac{2}{\omega\mu\sigma}}. \tag{3.25}$$

The loss resistance of the single loop antenna is expressed in this case as

$$R_\Omega(\omega) \approx \frac{2r}{\sigma(2a - \delta)\delta},$$ (3.26)

where we require $a > \delta$. For frequencies yielding a very pronounced skin effect, i.e. $\delta \ll a$, the loss resistance in (3.26) is further simplified as

$$R_\Omega(\omega) \approx \frac{r}{a}\sqrt{\frac{\omega\mu}{2\sigma}}$$ (3.27)

However, the loss resistance of a multi-turn loop antenna is not simply the result of (3.26) multiplied with the number of windings ν. The reason for this is that the charge in a tightly-wound conductor distorts the current distribution in the wire cross-section in addition to the skin effect. This behavior, called proximity effect [145], leads to further ohmic losses. The total loss resistance of a loop antenna with ν windings is therefore given as

$$R_{\Omega,\text{tot}} = \nu R_\Omega + R_P,$$ (3.28)

where R_P is the additional resistance caused by the proximity effect. Calculations of its value for loop antenna geometries are found in [144].

In the MQS regime, any resistance due to radiation losses is typically orders of magnitude smaller than the ohmic resistance. The radiation resistance R_R of a loop antenna with ν windings can be calculated as [12]

$$R_R = \nu^2 \eta_0 \frac{\pi}{6}\left(\frac{2\pi r}{\lambda}\right)^4,$$ (3.29)

where $\eta_0 = \mu \|\mathbf{E}\|/\|\mathbf{B}\|$ is the free space wave impedance, and the assumption was used that the antenna is small compared to the wavelength λ of interest. To illustrate the ratio of ohmic and radiative losses, we consider a single-turn loop antenna made of copper wire ($\sigma = 5.96 \cdot 10^7$ S/m) with a radius of $r = 2.5$ cm and a wire radius of $a = 250\,\mu$m. At a frequency of 10 MHz, the ohmic resistance is calculated as $R_\Omega = 8.1 \cdot 10^{-2}\,\Omega$, while the radiation resistance is $R_R = 1.5 \cdot 10^{-7}\,\Omega$. We therefore choose to neglect the radiation resistance in the following.

In the context of the previously discussed N-port network model for inductively coupled antennas, all loss mechanisms of the nth antenna can be summarized in a resistance R_n, which is placed in series with its corresponding inductance L_n [12]. We can then construct the *coupling impedance matrix* $\mathbf{Z}_\mathrm{C}$, describing N inductively coupled, lossy antennas, as

$$\mathbf{Z}_\mathrm{C} = \mathrm{diag}\left\{R_1, \ldots, R_N\right\} + j\omega \begin{bmatrix} L_1 & \cdots & M_{1N} \\ \vdots & \ddots & \vdots \\ M_{N1} & \cdots & L_N \end{bmatrix}, \tag{3.30}$$

where we write $\mathrm{diag}\left\{\cdot\right\}$ for a diagonal matrix with the elements of the argument. In the following section, we will use this matrix as the central element of a circuit-theoretic system description of inductively coupled nodes.

3.2 Circuit Theoretic System Description

This section will derive a formal framework for the description of signal and noise components in an inductively coupled network. We will base this description on multiport circuit theory. It has been noted in literature that a circuit theoretic system description for general wireless communication systems presents a suitable bridge between an accurate but computationally complex system description in terms of Maxwell's equations, and an abstract system description in information theoretic terms [132], [68].

3.2.1 Load Modulation

Before formulating a generalized communication model, we provide a brief overview of one of the most commonly used modulation schemes in inductively coupled communication systems, namely *load modulation*, which can be understood using a circuit theoretic approach. To this end we consider an inductively coupled system in which a sensor node transmits data to a central reader device. The basic idea of load modulation is that the sensor node may change the impedance of the circuit connected to its antenna in order to transmit information.

Fig. 3.5 shows an exemplary circuit implementing load modulation. The reader uses a source with voltage u_S and internal resistance R_S to supply power to the sensor node. The nodes employ a simple but commonly used design for impedance matching consisting for the capacitances C_1 and C_2, which achieve resonant behavior of reader and sensor node at a frequency ω_res, thus maximizing the effective power transfer over the inductive link [132]. The resistance R_L represents the input impedance of the circuitry implementing the power supply of the sensor node logic.

The load modulation at the sensor node is achieved by closing the switch of the modulation impedance Z_mod. The data can be detected at the reader as modulation of the complex

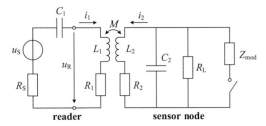

Figure 3.5. Circuit model for communication from sensor node to reader using load modulation.

amplitude of the antenna voltage u_R, which is calculated as

$$u_R = i_1 (R_1 + j\omega L_1) + i_2 \cdot j\omega M. \tag{3.31}$$

Herein the values of i_1 and i_2 depend on the current state of the modulating switch. Their values are obtained as

$$\begin{bmatrix} i_1 \\ i_2 \end{bmatrix} = \left(j\omega \mathbf{Z}_C + \begin{bmatrix} R_S + \frac{1}{j\omega C_1} & 0 \\ 0 & \left(\frac{1}{R_L} + j\omega C_2 + Y_{mod} \right)^{-1} \end{bmatrix} \right)^{-1} \begin{bmatrix} u_S \\ 0 \end{bmatrix} \tag{3.32}$$

were $Y_{mod} = \frac{1}{Z_{mod}}$ if the switch is closed, and $Y_{mod} = 0$ if the switch is open. Furthermore $\mathbf{Z}_C$ is the coupling impedance matrix defined in the previous section.

3.2.2 SISO Communication Model

Although the principle of communication based on load modulation is widely used in RFID systems due to its simplicity, an analysis of the communication performance of inductively coupled sensor networks calls for a more general model for several reasons. Most importantly, restricting the communication to the described modulation form imposes an arbitrary constraint. In addition, analysis of load modulation in a communication context is not straightforward due to the time-variant nature of the underlying network, as discussed in Section 2.5. Finally the provided description does not account for any secondary nodes present in the system. It is therefore desirable to have system model allowing an arbitrary number of transmitters, receivers, and secondary nodes, as well as a generic modulation scheme. In [68], a generalized circuit model for multipoint-to-multipoint wireless communication systems is presented. We will use this approach as basis for the development of a circuit model for inductively coupled systems. Specifically, we consider a single-input, single-output (SISO) communication system, i.e. the case with a single transmitter and single receiver. An arbitrary number of secondary nodes, which are neither transmitter nor receiver, may be present and through this presence influence the communication. Fig. 3.6 shows the circuit model for a SISO system consisting only of the transmitting and receiving node. Signal generation at the

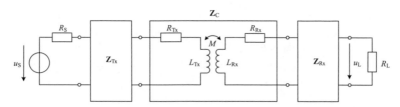

Figure 3.6. SISO system model for inductively coupled communication.

transmitter is modeled by a generic source with voltage u_S and internal source resistance R_S. At the receiver node, the received signal is given by the voltage u_L across the load resistance R_L modeling the input resistance of a low-noise amplifier (LNA). To optimize signal transfer between the signal source and load, both the transmitter and the receiver employ *matching networks* which are connected to the respective ports of the impedance coupling matrix $\mathbf{Z}_C$. For each frequency ω they can be described as the impedance matrices $\mathbf{Z}_{Tx}$ and $\mathbf{Z}_{Rx}$. The matching networks may consist of arbitrary passive circuit elements, implying that the complexity of their circuits increases with the bandwidth over which the matching is performed. Typically matching networks are realized as lossless networks which we will also assume in the following, i.e. $\mathfrak{Re}\left\{\mathbf{Z}_{Tx}\right\} = \mathfrak{Re}\left\{\mathbf{Z}_{Rx}\right\} = \mathbf{0}$ (with $\mathbf{0}$ denoting the zero matrix).

We are interested in expressing the received signal voltage u_L in terms of the source voltage u_S. This relation can be found by repeated formulation of the Thévenin equivalent of the network. In a first step, the Thevenin equivalent of the source and the transmitter matching network is formed, which is given by a source having an internal impedance identical to the output impedance[2] $Z_{Tx,out}$ of the transmitter matching network (cf. (2.18)) and the equivalent source voltage u_S' which follows from impedance parameter representation of the matching network as

$$u_S' = \frac{Z_{Tx,21}}{R_S + Z_{Tx,11}} \cdot u_S. \tag{3.33}$$

---

[2]In Fig. 3.6 we define the left-hand ports of all networks as input ports, and the right-hand ports as output ports.

By iteratively repeating the Thévenin reduction, the relation between source voltage and receiver voltage is obtained:

$$u_L = \underbrace{\frac{R_L}{R_L + Z_{Rx,out}}}_{G_L} \cdot \underbrace{\frac{Z_{Rx,21}}{Z_{Rx,11} + Z_{C,out}}}_{G_{Rx}} \cdot \underbrace{\frac{Z_{C,21}}{Z_{C,11} + Z_{Tx,out}}}_{G_C} \cdot \underbrace{\frac{Z_{Tx,21}}{Z_{Tx,11} + R_S}}_{G_{Tx}} \cdot u_S. \tag{3.34}$$

Here, the multiplicative terms of the complex voltage gain can be intuitively summarized as the factors G_{Tx}, G_C, G_{Rx}, and G_L, describing the individual gains of the source voltage past the transmitter matching network, coupling network, and receiver matching network, as well as the load resistor voltage divider, respectively. The total voltage gain can therefore be written as $G_{tot} = G_{Tx} \cdot G_C \cdot G_{Rx} \cdot G_L$.

To accurately model the noise affecting the communication, all possible noise sources must be taken into account. Fig. 3.7 shows the various locations at which noise is generated within the communication system. The origins of noise are the following:

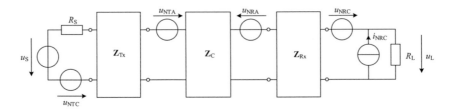

Figure 3.7. SISO system model of inductively coupled communication including noise sources.

Transmit circuit noise The voltage u_{NTC} represents thermal noise stemming from the internal resistance of the signal source. The actual source voltage u_S is assumed to be noiseless. It should be noted that the transmit-side matching network has no noise contribution due to the assumption of losslessness. Following the discussion in Section 2.4, the noise variance is calculated as $E\left[|u_{NTC}|^2\right] = \sigma_{NTC}^2 = 4k_B T \Delta f R_S$. Herein k_B is the Boltzmann constant, T is the temperature of the circuit, and Δf is the bandwidth of interest.

Transmit antenna noise As has been discussed in the previous section, the radiation resistance at the antenna of the transmitter is negligible at the frequencies of interest. Therefore the only noise contribution of the transmit antenna is the thermal noise voltage u_{NTA} stemming from the ohmic loss resistance R_{Tx}. Its variance is calculated as

$$\mathsf{E}\left[\left|u_{\mathrm{NTA}}\right|^2\right] = \sigma_{\mathrm{NTA}}^2 = 4k_{\mathrm{B}}T\Delta f R_{\mathrm{Tx}}.$$

Receive antenna noise Following the reasoning for the transmit antenna noise, the variance of the receive antenna noise is given by $\mathsf{E}\left[\left|u_{\mathrm{NTR}}\right|^2\right] = \sigma_{\mathrm{NTR}}^2 = 4k_{\mathrm{B}}T\Delta f R_{\mathrm{Rx}}$.

Receive circuit noise The resistance R_{L} models the input impedance of the LNA stage and all subsequent circuitry in the receiver chain. We consider only the noise of the LNA itself, which is a noisy, active two-port. As such, its noise contributions are modeled by one noise source per port, or equivalently two sources at the input port, which pertain to the voltage u_{NRC} and the current i_{NRC}. As u_{NRC} and i_{NRC} are generated in the same electrical network, they are generally correlated with correlation coefficient ρ_{C}. We make the definitions $\mathsf{E}\left[\left|u_{\mathrm{NRC}}\right|^2\right] = \sigma_{\mathrm{NRC}}^2$ and, following the notation in [68], $\mathsf{E}\left[\left|i_{\mathrm{NRC}}\right|^2\right] = \sigma_{\mathrm{NRC}}^2 / R_{\mathrm{N}}^2$. The factor R_{N} is called noise resistance. As the values of σ_{NRC}^2, R_{N}, and ρ_{C} depend on the implementation of the LNA, they must be specified using the corresponding data sheet or from experience.

The total noise voltage u_{LN} at the load resistance can be found by applying the respective gains defined in (3.34) to the individual noise contributions. We therefore obtain the relations

$$u_{\mathrm{LN}} = G_{\mathrm{L}}\left(Z_{\mathrm{Rx,out}} i_{\mathrm{NRC}} - u_{\mathrm{NRC}}\right) + G_{\mathrm{L}}G_{\mathrm{Rx}}u_{\mathrm{NRA}} - G_{\mathrm{L}}G_{\mathrm{Rx}}G_{\mathrm{C}}u_{\mathrm{NTA}} + G_{\mathrm{tot}}u_{\mathrm{NTC}} \qquad (3.35)$$

for the noise voltage and

$$\mathsf{E}\left[\left|u_{\mathrm{LN}}\right|^2\right] = \left|G_{\mathrm{L}}\right|^2\sigma_{\mathrm{NRC}}^2\left(1 + \frac{\left|Z_{\mathrm{Rx,out}}\right|^2}{R_{\mathrm{N}}^2} - 2\frac{\mathfrak{Re}\left\{Z_{\mathrm{Rx,out}} \cdot \rho_{\mathrm{C}}^*\right\}}{R_{\mathrm{N}}}\right)$$
$$+ \left|G_{\mathrm{L}}G_{\mathrm{Rx}}\right|^2\sigma_{\mathrm{NRA}}^2 + \left|G_{\mathrm{L}}G_{\mathrm{Rx}}G_{\mathrm{C}}\right|^2\sigma_{\mathrm{NTA}}^2 + \left|G_{\mathrm{tot}}\right|^2\sigma_{\mathrm{NTC}}^2. \qquad (3.36)$$

for the noise variance.

The formalization of signal and noise voltages at the receiver load allows for the analysis of the communication performance. This performance is affected by the choice of matching networks, where the particular matching network implementations can be optimized with respect to any desired property. A common and sensible choice is to maximize the available power from the source at the output of the transmitter-side matching network and maximize the signal-to-noise ratio (SNR) at the output of the receiver. General conditions on $\mathbf{Z}_{\mathrm{Tx}}$ and $\mathbf{Z}_{\mathrm{Rx}}$ achieving these properties are derived in [68] as

$$\mathbf{Z}_{\mathrm{Tx}} = \begin{bmatrix} 0 & -j\sqrt{R_{\mathrm{S}} \cdot \mathfrak{Re}\left\{Z_{\mathrm{C,in}}\right\}} \\ -j\sqrt{R_{\mathrm{S}} \cdot \mathfrak{Re}\left\{Z_{\mathrm{C,in}}\right\}} & -j\mathfrak{Im}\left\{Z_{\mathrm{C,in}}\right\} \end{bmatrix}, \text{ and} \qquad (3.37)$$

$$\mathbf{Z}_{\text{Rx}} = \begin{bmatrix} -j\Im\left\{Z_{\text{C,out}}\right\} & j\sqrt{\Re\left\{Z_{\text{C,out}}\right\} \cdot \Re\left\{Z_{\text{opt}}\right\}} \\ \sqrt{\Re\left\{Z_{\text{C,out}}\right\} \cdot \Re\left\{Z_{\text{opt}}\right\}} & j\Im\left\{Z_{\text{opt}}\right\} \end{bmatrix}. \tag{3.38}$$

Herein $Z_{\text{C,in}}$ and $Z_{\text{C,out}}$ are the input and output impedances of $\mathbf{Z}_{\text{C}}$, and Z_{opt} is defined as

$$Z_{\text{opt}} = R_{\text{N}}\left(\sqrt{1 - \Im\left\{\rho_{\text{C}}\right\}^2} + j\Im\left\{\rho_{\text{C}}\right\}\right). \tag{3.39}$$

In the following we will consider the SNR of the receive symbol as a figure of merit. To this end we assume that the nodes of an inductively coupled wireless sensor network move slowly or not at all, such that the wireless channel can be sensibly assumed as time-invariant. The symbol-discrete input-output relation of a the channel in this setting is given as

$$y = hx + n, \tag{3.40}$$

with $y, h, x, n \in \mathbb{C}$. Here x is the transmitted symbol, y is the received symbol, h is the channel gain, and n is the noise of the receiver. The SNR is defined in an information-theoretic context as

$$\text{SNR} = \frac{\mathsf{E}\left[|hx|^2\right]}{\mathsf{E}\left[|n|^2\right]}. \tag{3.41}$$

It has been noted in literature that the above definition, although being employed in many contributions in the field of information theory, is not based on physical units of power. The work in [68] addresses this issue in great detail, and provides a formulation to relate the previously introduced physical quantities of the communication system to the well-established information-theoretic framework by equating $\mathsf{E}\left[|x|^2\right]$ to the active power flowing into $Z_{\text{Tx,in}}$, as well as $\mathsf{E}\left[|y|^2\right]$ and $\mathsf{E}\left[|n|^2\right]$ to the power dissipated by the received signal voltage u_{L} and noise voltage u_{NL}, respectively, over R_{L}. This implies the relations

$$x = \sqrt{\frac{\Re\left\{Z_{\text{Tx,in}}\right\}}{|R_{\text{S}} + Z_{\text{Tx,in}}|^2}} \cdot u_{\text{S}}, \tag{3.42}$$

$$y = \sqrt{\frac{1}{R_{\text{L}}}} \cdot u_{\text{L}}, \tag{3.43}$$

$$n = \sqrt{\frac{1}{R_{\text{L}}}} \cdot u_{\text{LN}}, \text{ and} \tag{3.44}$$

$$h = \sqrt{\frac{1}{R_{\text{L}}} \cdot \frac{|R_{\text{S}} + Z_{\text{Tx,in}}|^2}{\Re\left\{Z_{\text{Tx,in}}\right\}}} \cdot G_{\text{tot}}. \tag{3.45}$$

3.2.3 Communication in the Presence of Secondary Nodes

For sensor networks with a physical layer based on far-field propagating waves the individual nodes are commonly placed far enough from each other to be effectively decoupled, meaning that secondary nodes (i.e. operating neither as transmitter or receiver) do not affect the communication by their physical presence. In contrast, in an inductively coupled network many nodes can be placed closely together due to the necessity of being within near-field communication range. The mutual inductive coupling of all involved antennas impacts both the desired signal and noise at the receiving nodes, even if the secondary nodes are completely passive. As will be discussed in Chapters 4 and 5, this interaction can introduce significant benefits to the communication similar to those achievable by wireless relaying in far-field systems. For this reason secondary nodes may also be referred to as relay nodes, particularly if their presence is intentional.

To illustrate the interaction, Fig. 3.8 shows the previously studied SISO communication system in the presence of N_S secondary nodes. We assume that the antennas of all sec-

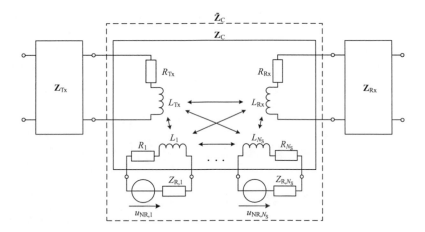

Figure 3.8. SISO system model in the presence of N_S passive secondary nodes with their noise sources shown.

ondary nodes are terminated by load impedances[3] $Z_{R,m}$ with $m \in \{1, \ldots, N_S\}$. The coupling impedance matrix $\mathbf{Z}_C$ takes the N-port form given in (3.30), where $N = N_S + 2$ in the SISO case. To account for the presence of the secondary nodes on the SISO communication link, they can be encapsulated by a modified two-port coupling impedance matrix $\tilde{\mathbf{Z}}_C$. To this end we utilize the notion of a partially terminated impedance matrix as discussed in Section 2.2. In the specific case of the coupling impedance matrix $\mathbf{Z}_C$ being partially terminated by the loads of the secondary nodes there are no transimpedance terms between the terminations of the individual nodes. This means that the load matrix—which we in this setup denote as $\mathbf{Z}_R$—is therefore given as

$$\mathbf{Z}_R = \mathrm{diag}\left\{Z_{R,1}, \ldots, Z_{R,N_S}\right\}, \tag{3.46}$$

and an equivalent two-port coupling impedance matrix $\tilde{\mathbf{Z}}_C$ can be obtained using (2.16).

The secondary nodes also introduce additional noise contributions. This noise is, as for the previously discussed setup, predominantly thermal noise stemming from the antenna's ohmic loss resistance as well as any real part of the loads $Z_{R,m}$. Each secondary node therefore introduces a noise source as depicted in Fig. 3.8 with the noise voltage $u_{NR,m}$ and respective variance $\mathsf{E}\left[|u_{NR,m}|^2\right] = \sigma_{NR,m}^2 = 4k_B T \Delta f \left(R_m + \mathfrak{Re}(Z_{R,m})\right)$. In consequence of the linearity of the overall system, the joint noise contribution of all secondary nodes result in an additive term to the total noise voltage and variance in (3.35) and (3.36). To evaluate the noise propagating from the secondary nodes to the receiver node, we denote by $\mathbf{Z}_{C,Tx}$ the impedance matrix resulting from the partial termination of $\mathbf{Z}_C$ with $Z_{Tx,out}$ at the transmitter port. It follows from the symmetry of the coupling impedance matrix that $\mathbf{Z}_{C,Tx}$ has the form

$$\mathbf{Z}_{C,Tx} = \begin{bmatrix} s & \mathbf{t}^T \\ \mathbf{t} & \mathbf{U} \end{bmatrix}. \tag{3.47}$$

Using this notation it is possible to construct a Thévenin equivalent source of all secondary node noise contributions with the equivalent voltage [33]

$$u'_{NR} = \mathbf{t}^T \left(\mathbf{U} + \mathbf{Z}_R\right)^{-1} \mathbf{u}_{NR}. \tag{3.48}$$

Here the vector $\mathbf{u}_{NR} = [u_{NR,1}, \ldots, u_{NR,N_S}]^T$ contains the noise voltages at all secondary nodes. The total voltage u'_{NR} in (3.48) occurs at the input port of the receiver matching network and is therefore transformed to the load resistance with a gain of $G_L \cdot G_{Rx}$ following the same

---

[3]Here, both the matching network and termination impedances of the secondary nodes are encapsulated in the loads $Z_{R,m}$.

calculation as for the previously introduced noise contribution u_{NRA} (cf. Fig. 3.7). The total noise voltage at the load resistance from all contributions in the presence of relays is therefore given as

$$\tilde{u}_{\mathrm{LN}} = u_{\mathrm{LN}} + G_{\mathrm{L}} \cdot G_{\mathrm{Rx}} \cdot \mathbf{t}^{\mathrm{T}} \left(\mathbf{U} + \mathbf{Z}_{\mathrm{R}} \right)^{-1} \mathbf{u}_{\mathrm{NR}}, \tag{3.49}$$

with u_{LN} given in (3.35). Equivalently, the total noise variance is calculated as

$$\begin{aligned}
\mathsf{E}\left[|\tilde{u}_{\mathrm{LN}}|^2 \right] &= \mathsf{E}\left[|u_{\mathrm{LN}}|^2 \right] \\
&\quad + |G_{\mathrm{L}} \cdot G_{\mathrm{Rx}}|^2 \cdot \mathbf{t}^{\mathrm{T}} \left(\mathbf{U} + \mathbf{Z}_{\mathrm{R}} \right)^{-1} \cdot \mathbf{\Sigma}_{\mathrm{NR}} \cdot \left(\mathbf{t}^{\mathrm{T}} \left(\mathbf{U} + \mathbf{Z}_{\mathrm{R}} \right)^{-1} \right)^{\mathrm{H}},
\end{aligned} \tag{3.50}$$

where $\mathbf{\Sigma}_{\mathrm{NR}} = \mathrm{diag}\left\{ \sigma_{\mathrm{NR},1}^2, \ldots, \sigma_{\mathrm{NR},N_{\mathrm{S}}}^2 \right\}$ is the covariance matrix of the noise voltages at the secondary nodes, and $\mathsf{E}\left[|u_{\mathrm{LN}}|^2 \right]$ is defined in (3.36).

3.3 Descriptive Limitations and Approximations

The presented descriptions of inductively coupled communication systems have been primarily derived to describe applications such as RFID and NFC, which use antenna dimensions on the order of centimeters and operating frequencies in the kHz to low MHz range. To judge the feasibility of inductively coupled microsensor networks it is important to understand both the limitations and scaling behavior of the underlying models.

3.3.1 Valid Frequency Range for MQS Approximation

We have used the MQS approximation in the previous sections to obtain a simple description of a system of inductively coupled antennas. It is sensible to investigate the validity of the underlying assumption of sufficiently slowly changing currents within the operational parameters of interest in this work. A truly static current density J induces a static magnetic flux density $\mathbf{B}$ as discussed in Section 3.1. In a quasi-static setting, however, the current density and therefore the $\mathbf{B}$-field are slowly changing. According to Faraday's law (cf. (3.1)), this induces an $\mathbf{E}$-field which in turn is the source of a magnetic flux density $\tilde{\mathbf{B}}$. The additional field $\tilde{\mathbf{B}}$, which according to Lenz's law opposes the original, quasi-static field $\mathbf{B}$, is not accounted for in the MQS description of the system. Therefore the ratio $\|\tilde{\mathbf{B}}\|/\|\mathbf{B}\|$ must be small for the MQS approximation to be valid.

The work in [53] provides an analysis of this condition for conductor arrangements whose geometric dimensions—e.g. length, diameter, thickness, etc.—are all on the same order of magnitude[4]. The size of the conductor arrangement is then described by a single characteristic length D. Assuming a sinusoidal current, the integral form of Faraday's law can then be approximated in the MQS regime as

$$\oint_{\delta\Sigma} \mathbf{E} \cdot d\mathbf{l} = -\int_{\Sigma} \frac{\partial \mathbf{B}}{\partial t} da$$
$$\approx \|\mathbf{E}\|\, D \approx -j\omega \|\mathbf{B}\|\, D^2. \tag{3.51}$$

The error field $\tilde{\mathbf{B}}$ is characterized similarly by evaluating Ampères law (3.2) for the originally neglected displacement current:

$$\oint_{\delta\Sigma} \tilde{\mathbf{B}} \cdot d\mathbf{l} = \int_{\Sigma} \mu\varepsilon \frac{\partial \mathbf{E}}{\partial t} da$$
$$\approx \|\tilde{\mathbf{B}}\|D \approx j\omega\mu\varepsilon \|\mathbf{E}\|\, D^2. \tag{3.52}$$

Subsituting (3.51) into (3.52) we find the ratio

$$\frac{\|\tilde{\mathbf{B}}\|}{\|\mathbf{B}\|} = \omega^2 \mu\varepsilon D^2, \tag{3.53}$$

which establishes the condition $\omega^2\mu\varepsilon D^2 \ll 1$ for the MQS approximation to be justified.

In a system of inductively coupled nodes, however, the typical size scales of the antenna wire thickness, the coil radius, and the distance between the antennas may be vastly different. To evaluate the error field ratio $\|\tilde{\mathbf{B}}\|/\|\mathbf{B}\|$ in this system, we need to consider the magnetic field of a loop antenna both with and without the MQS approximation. For an electrically small thin-wire loop antenna located at the origin with radius r, windings ν, and current i, the MQS approximation of the magnetic flux density $\mathbf{B}$ is stated in (3.18). The full-wave solution of the magnetic field generated by such an antenna can be developed by approximating the magnetic vector potential as a truncated series expansion as shown in [12]. The resulting magnetic flux density $\mathbf{B}_{\text{fw}}$ is given in spherical coordinates as

$$\mathbf{B}_{\text{fw}} = \begin{bmatrix} B_\rho \\ B_\vartheta \\ B_\varphi \end{bmatrix} \approx \mu\nu i \begin{bmatrix} \frac{j2\pi r^2 \cos\vartheta}{2\lambda\rho^2}\left(1 + \frac{\lambda}{j2\pi\rho}\right) \\ -\frac{(\pi r)^2 \sin\vartheta}{\lambda^2\rho}\left(1 + \frac{\lambda}{j2\pi\rho} - \frac{\lambda^2}{(2\pi\rho)^2}\right) \\ 0 \end{bmatrix} \cdot e^{-j\frac{2\pi}{\lambda}\rho}, \tag{3.54}$$

---

[4]This condition is defined by the authors as all dimensions being within a factor of 2 of each other.

where λ is the wavelength of the exciting current. It can be noted that at large separations, i.e. for $\rho \to \infty$, the far-field terms with magnitude proportional to ρ^{-1} are dominant. Likewise for $\rho \to 0$ only the near-field components proportional to ρ^{-3} remain, and the magnitude of the expression in (3.54) becomes identical to the MQS solution in (3.18).

By defining the error field as

$$\tilde{\mathbf{B}} = \mathbf{B}_{\mathrm{fw}} - \mathbf{B} \qquad (3.55)$$

where $\mathbf{B}$ is the MQS approximation given in (3.18), we calculate the error field ratio $\|\tilde{\mathbf{B}}\|/\|\mathbf{B}\|$ for a set of frequencies and distances of interest for inductively coupled sensor networks in the macroscale and microscale. The results are visualized in Fig. 3.9 with the field source being an antenna with $r = 2.5\,\mathrm{cm}$ and $\nu = 1$. The fields were evaluated for an elevation angle of $\pi/2$, corresponding to the xy-plane in Cartesian coordinates.

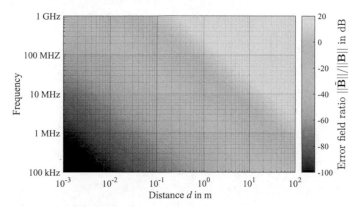

Figure 3.9. Error field ratio $\|\tilde{\mathbf{B}}\|/\|\mathbf{B}\|$ of a loop antenna with $r = 2.5\,\mathrm{cm}$, $\nu = 1$ for different frequencies and distances.

The results show a tradeoff between the distance d and the maximum frequency ω of an exciting current for which the MQS approximation is valid, i.e. $\|\tilde{\mathbf{B}}\|/\|\mathbf{B}\| \ll 1$. At the frequency of 25 MHz, which will often be used throughout the following chapters, the MQS regime therefore applies to node separations of up to a few meters. At greater distances, the observed magnetic field is dominated by the far-field propagating waves radiated by the source. The far-field does not retroact on the source and therefore represents a mechanism of power and information transfer that is distinct from inductive coupling.

51

3.3.2 Phase Relation of Inductively Coupled Antennas

Besides the presence of far-field propagating waves, the full-wave solution of the magnetic flux density in (3.54) shows that both the propagating and the evanescent near-field components exhibit a non-zero phase. In contrast we have assumed under the MQS approximation that all currents in the communication system change slowly enough to be approximately static. Following this assumption, a change in a current i instantly results in a change in the magnetic field $\mathbf{B}(i)$ it excites. This implies that the communication system has no delay and any phase shift between the source voltage at a transmitting node and the load voltage at a receiver is independent of the relative position of the devices. A delay-free channel would have interesting consequences in a communications context. For example multiple synchronous transmitters could always achieve coherent superposition of their signals at a receiver without requiring channel state information (CSI). However, as all electromagnetic fields propagate from changes in their sources with a speed-of-light delay, we expect a distance-dependent phase shift of the received signal. The question remains at which frequencies this effect becomes visible.

To evaluate the phase relationship between a transmitter and a receiver antenna, we consider an ideal inductively coupled two-port, i.e. lossless coupled antennas without matching networks. We are interested in the phase shift $\Delta\phi$ between the voltages u_m at the antenna port of the transmitter and u_n at the port of the receiver. For an arbitrary two-port described by an impedance parameter matrix $\mathbf{Z}$, the phase shift $\Delta\phi$ can be calculated as [14]

$$\Delta\phi = \arg\left\{\frac{u_n}{u_m}\right\} = \arg\left\{\frac{[\mathbf{Z}]_{2,1}}{[\mathbf{Z}]_{1,1}}\right\}, \tag{3.56}$$

assuming the receiver port is open circuited. In the case of coupled loop antennas, the elements of the impedance matrix follow from (3.10) as $[\mathbf{Z}]_{2,1} = j\omega M_{mn}$ and $[\mathbf{Z}]_{1,1} = j\omega L_m$.

We obtain M_{mn} by inserting (3.54) into (3.19) and (3.8), i.e. we again use the assumption of the magnetic field being constant over the receive antenna surface $\mathbf{a}_n$. For this general case the mutual inductance M_{mn} becomes a complex number due to the phase term in (3.54). To evaluate the phase relationship we assume that the considered antenna pair is placed in a 2D coplanar placement arrangement with distance d between the antenna centers. Fig. 3.10 shows the resulting phase $\Delta\phi$ between u_n and u_m in degrees, as the distance is varied from 5 cm to 1.6 m.

The phase shift for the low frequency of 100 kHz corresponds to the value of 0° independent of the distance in the range of interest, which is the expected value for an ideal transformer circuit. In fact there is almost no observable distance dependence even at a frequency of 10 MHz.

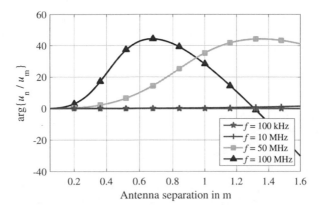

Figure 3.10. Comparison phase shift over distance for different frequencies. Both antennas have a radius of $r = 2.5\,\text{cm}$ and $\nu = 1$ winding.

However, the results for $50\,\text{MHz}$ and $100\,\text{MHz}$ show significant variation of the phase shift with increasing variation. It should be noted that, in practice, the effective communication range of the considered system is well below one meter as will be shown in following parts of this work, such that the large phase shifts at greater difference might be irrelevant for many systems. On the other hand, in the context of range extension methods studied in Chapter 4, this effect might lead to previously unconsidered challenges and opportunities. For example, deliberate manipulation of the phase shifts in a multiuser communication setup could be used to optimize the signal-to-interference-plus-noise ratio (SINR) using beamforming methods such as zero-forcing [161].

3.4 Scaling Behavior of Miniaturized Inductively Coupled Networks

The low complexity requirements and inherent ability for external power supply have motivated us to identify inductively coupled communication as a viable technology to implement wireless communication among highly miniaturized sensor networks, as discussed in the previous chapter. To provide intuition on the behavior of inductive coupling as node sizes are decreased, it is useful to identify the physical scaling laws (cf. [168]) of the communication system. To this end we scale down all sizes in an inductively coupled network by a factor of l. By evaluating how the individual physical quantities involved in the description of the network scale with l, several insights into the overall behavior of miniaturized networks can be gained.

3.4.1 Self- and Mutual Inductance

To estimate the scaling behavior of the self-inductance L for a loop antenna with ν windings, we approximate its value as

$$L \approx \nu^2 L_{\text{single}}, \tag{3.57}$$

where L_{single} is the self-inductance of a single circular loop with radius r and wire radius a. For this geometry, the general inductance expression in (3.17) has the solution

$$L_{\text{single}} = \mu r \cdot \left(\ln \left(8\frac{r}{a} \right) - 2 + \Upsilon/2 \right) + \ldots, \tag{3.58}$$

where Υ is defined in Section 3.1. The error of this expression due to neglecting the omitted terms decays with $(a/r)^2$ [29]. It follows that as $r \propto l$, the self-inductance of the antenna scales as $L \propto l$.

The dipole approximation of the mutual inductance in (3.20) is proportional to r^4/d^3. With $d \propto l$ we obtain $M \propto l$. To show that this scaling holds not only for the dipole approximation but for arbitrary antenna arrangements, we consider the Neumann integral in (3.15). A linear scaling of the node antenna arrangement corresponds to the modified integrand

$$\frac{\mathrm{d}(l\mathbf{l}_m) \cdot \mathrm{d}(l\mathbf{l}_n)}{\|l\mathbf{l}_m - l\mathbf{l}_n\|} = \frac{l^2}{l} \cdot \frac{\mathrm{d}\mathbf{l}_m \cdot \mathrm{d}\mathbf{l}_n}{\|\mathbf{l}_m - \mathbf{l}_n\|}, \tag{3.59}$$

which verifies the result.

3.4.2 Loss Resistance

The losses of the node antennas are modeled by the ohmic loss resistance R_Ω and radiative loss resistance R_{R}. The ohmic loss resistance of a multi-turn loop is calculated according to (3.28). As the proximity effect is invariant to l (cf. [12]), the scaling behavior of R_Ω only depends on the single-turn loss resistance given in (3.26) which in turn depends on the ratio of wire radius and skin depth a/δ. The skin depth δ is a function of frequency. For low frequencies, the loss resistance is equivalent to (3.24) which has a scaling behavior of r/a^2

and therefore $R_\Omega \propto l/l^2 = l^{-1}$. On the other hand, for high frequencies the loss resistance expression is approximated by (3.27). It scales as r/a, and accordingly $R_\Omega \propto l^0$, i.e. the ohmic loss resistance does not scale with l. The radiation resistance in (3.29), on the other hand, scales as $R_R \propto l^4$ and therefore becomes increasingly negligible when considering node miniaturization, as pointed out in Section 3.1.

To illustrate the regimes of the different scaling laws for R_Ω, we consider the condition $a = 10\delta$ as border between them. Fig. 3.11 visualizes the regimes for an exemplary loop antenna made of copper with $r/a = 100$ as a function of frequency. The location of the

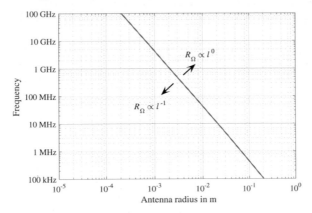

Figure 3.11. Scaling behavior of ohmic loss resistance R_Ω.

border indicates that for the typical frequencies used for inductive coupling, i.e. the kHz to MHz range, the crossover between the scaling regimes takes places for nodes on the order of centimeters to millimeters. In nodes of extremely small size the increase in loss resistance can therefore be expected to impose prohibitive design requirements.

3.4.3 Communication Performance

We evaluate the communication performance as a function of the scaling factor using the SISO circuit model introduced in Section 3.2. A suitable measure of communication performance is the SNR as defined in (3.41). Typically the operation of a communication system takes place in the loosely coupled regime, i.e. $|k| \ll 1$. In this regime the receiver noise sources u_NRC and i_NRC can be considered as the dominant noise sources, implying that the SNR primarily depends on the channel gain $|h|^2$.

We consider a resonantly matched pair of antennas antennas denoted by indices m and n transmitting a narrowband signal at resonance frequency. It can be shown that for optimally chosen matching networks, the maximal value of $|h|^2$ can be calculated as [83]

$$|h|^2_{\max} = \frac{k^2 \omega^2 \frac{L_m L_n}{R_m R_n}}{\left(1 + \sqrt{1 + k^2 \omega^2 \frac{L_m L_n}{R_m R_n}}\right)^2} \overset{k \ll 1}{\approx} \frac{k^2 \omega^2 L_m L_n}{4 R_m R_n}. \tag{3.60}$$

As the expression for $|h|^2_{\max}$ depends on R_m and R_n, its scaling behavior is subject to the previously identified regimes for resistance scaling. With the coupling coefficient $k_{mn} = M_{mn}/\sqrt{L_m L_n}$ being independent of l as $k_{mn} \propto l/l = l^0$, we find—under the assumption of loose coupling—that $|h|^2_{\max} \propto l^2$ for comparatively large antennas in the regime of $R_\Omega \propto l^0$, and $|h|^2_{\max} \propto l^4$ for antennas in the regime of $R_\Omega \propto l^{-1}$. Accordingly, decreasing the size of a communication system by a factor of 10 leads to an SNR loss of 20 dB or 40 dB, depending on the scaling regime of the antenna loss resistance. This behavior can intuitively be understood as a consequence of the reduced intrinsic quality factors $Q_m = \omega L_m/R_m$ of the antennas.

To numerically verify the behavior of the SNR, we investigate a coaxially arranged antenna pair as depicted in Fig. 3.3. The antennas have a radius of $r = l \cdot 2.5$ cm and $\nu = 5$ windings. We use fixed matching networks $\mathbf{Z}_{\text{Tx}}$ and $\mathbf{Z}_{\text{Rx}}$ which are independent of the antenna positions. This choice is suboptimal as the resulting SNR can be improved by choosing the matching adaptively for each antenna arrangement according to the conditions for SNR maximization in (3.37) and (3.38). On the other hand, practical implementations of microsensor networks are unlikely to use adaptive matching due to the increased complexity requirements. We design the fixed matching networks with the assumption of uncoupled nodes, which is asymptotically optimal for increasing distance. Additional simulation parameters are given in Table 3.1. Fig. 3.12 visualizes the obtained SNR over the distance normalized to the antenna radius, where the scaling factor of l was decreased from 1 to $1/1000$.

There are several observations to be made from these results. First, at separations significantly greater that the antenna radius r, the SNR drops off with a negative slope of 60 dB per decade. This is in line with the behavior of the magnetic near-field which has a distance dependency of $\mathbf{B} \propto d^{-3}$ for $d \gg r$, as follows from (3.18). Compared to wireless communication systems operating in the far-field, which exhibit a free space path loss of 20 dB per decade (cf. (6.1)), it becomes clear that a major limitation of inductively coupled sensor networks is the limited communication range of the nodes. Assuming an SNR value of 0 dB as the threshold for reliable communication, the usable ranges are found as 3.6 m for $r = 2.5$ cm, 15.1 cm for $r = 25$ mm, 3.5 mm for $r = 250\,\mu$m, and 68.3 μm for $r = 25\,\mu$m. The disproportional decrease

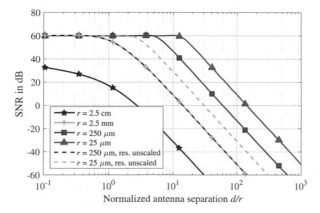

Figure 3.12. SNR scaling behavior over normalized antenna separation.

Parameter	Symbol	Value
Bandwidth	Δf	$1\,\text{kHz}$
Noise temperature	T	$300\,\text{K}$
Correlation coefficient	ρ_C	0
Transmit power	P_T	$\mathsf{E}\left[x^2\right] = 1\,\text{nW}$
Antenna conductivity	σ	$58.5 \cdot 10^6\,\text{Sm}^{-1}$
LNA noise figure	F	$15\,\text{dB}$
Receive circuit noise variance	σ^2_{NRC}	$4Fk_BT\Delta f R_L$
Source resistance	R_S	$5\,\Omega$
LNA input resistance	R_L	$50\,\Omega$
Noise resistance	R_N	$50\,\Omega$

Table 3.1. Simulation parameters for scaling analysis of SNR.

of usable range stems from the SNR scaling behavior. Going from $l = 1$ to $l = 1/10$, an SNR loss of approximately 20 dB can be observed in the loosely coupled regime, which is in line with the resistance scaling at the considered antenna sizes. As l is further decreased to $1/100$ and $1/1000$, yielding highly miniaturized antennas with $r = 250\,\mu$m and $r = 25\,\mu$m, the SNR loss becomes greater, as can be explained primarily by the increase in antenna loss resistance due to the wire diameter a approaching the size of the skin depth δ. To verify this conjecture, the dashed line again shows the SNR for the two smallest antenna sizes of $r = 250\,\mu$m and $r = 25\,\mu$m with the difference that the loss resistance R_Ω was chosen identical to the value for the unscaled antennas with $r = 2.5$ cm. As expected, the SNR in the loosely coupled regime then shows 20 dB loss for each scaling by a factor of $1/10$.

Part II.

Cooperative Communications in Inductively Coupled Networks

4

Range Extension by Relaying

A fundamental limitation of the communication among inductively coupled sensors lies in the short communication range, which is a result of the near-field nature of the coupling mechanism. Some applications, such as NFC electronic payment, exploit this limitation in range to increase privacy and security of the communication. However, the use case of data gathering typical for sensor network settings strongly benefits from high degrees of interconnectivity which imposes requirements on the communication range. In the previous chapter we have discussed that the SNR of an inductively coupled communication link drops off with the sixth power of the distance between the nodes, and an additional SNR degradation can be expected as the dimensions of the sensor network are reduced. Realizing a highly interconnected inductively coupled network is therefore challenging.

It is a well known approach in wireless communications to overcome the drawback of limited range by the use of relays, i.e. devices which, typically positioned between the source and destination, aid the communication. Besides the mere extension of range, the use of relaying has many distinct benefits. Particularly low-complexity sensor networks can benefit from improved energy efficiency [18]. Other benefits of relaying include the possibility of exploiting the spatial structure of the relays to achieve spatial diversity or spatial multiplexing.

Relay nodes have successfully been studied in many different communication systems, ranging from cellular networks [124] to local area networks [147]. In this chapter, we analyze the communication performance in inductively coupled networks using relays. To this end, we first provide an overview of possible relay implementations which can be categorized into active and passive approaches. Passive relay implementations are particularly promising for use in microsensor networks as they can be implemented as a simple resonant circuit without power supply or logic. We provide a quantitative analysis of the performance gains which can be achieved from employing passive relays in inductively coupled networks. In contrast to previous work, the underlying system model—presented in the previous chapter—allows for a flexible numerical analysis of arbitrary network configurations.

4.1 Classical Relaying

In wireless communications a relay node can be generally defined as a node that assists the communication by receiving a signal intended for another node, possibly modifying it, and forwarding, i.e. retransmitting it [87]. Established relaying approaches found in literature can be divided into two categories: regenerative and non-regenerative relaying.

Regenerative relaying The most common regenerative relaying approach is *decode-and-forward (DF) relaying*. Herein the relay attempts to decode the signal from the source. The decoded message is subsequently re-encoded (possibly using a different codebook), and retransmitted to the destination. If a DF relay can decode successfully it transmits an exact copy of the source message, thus regenerating the noisy received signal. On the other hand, erroneous decoding is irreversible at the intended destination. The achievable rate of a DF relaying link is therefore limited by the relays ability to decode. If the signal received by the relay is very weak, a *soft decode-and-forward relaying* approach may be used instead [146]. To this end the relay generates a message containing soft information about its received message, e.g. the log-likelihood ratios of the individual bits. This information can be utilized by the destination to decode the original message.

Non-regenerative relaying Non-regenerative relaying approaches only consider the received noisy signal instead of taking the represented message into account. The simplest non-regenerative relaying approach is *amplify-and-forward (AF) relaying*, in which the signal received from the source is amplified, i.e. multiplied with a (complex) scalar, and retransmitted to the destination. While the AF operation is of low complexity, it has the drawback of also amplifying any noise or interference received alongside the desired signal. The implementation of AF relaying in digital devices requires the quantization of the received signal. The resulting scheme is sometimes considered an approach in its own right, referred to as *quantize-and-forward (QF) relaying*. QF relaying can generally be improved by performing source coding at the relays. This approach is known in literature as *compress-and-forward (CF) relaying*.

Due to its low complexity AF relaying is a natural choice for small scale sensor networks. In a symbol discrete SISO system model, the signal y_{SR} received by the relay can be written as

$$y_{SR} = h_{SR}x + n, \tag{4.1}$$

where x is the symbol transmitted by the source, h_{SR} is the channel coefficient from the source to the relay, and n denotes the receiver noise at the relay. The received signal is multiplied by the scaling coefficient g and retransmitted, such that the destination receives the signal.

$$y_{\mathrm{RD}} = g \cdot h_{\mathrm{RD}}(h_{\mathrm{SD}}x + n) + m, \qquad (4.2)$$

with h_{RD} and m denoting the channel between relay and destination as well as the noise at the destination, respectively. A typical problem of designing AF relaying networks lies in the choice of the gain coefficients g or relay positions (which influence the channel coefficients) in order to optimize a chosen figure of merit. As an example, different gain allocation strategies to achieve cooperative diversity in fading channels using multiple relays are presented in [50].

4.2 Relaying in Inductively Coupled Networks

4.2.1 Active Relaying

The notion of AF relaying is easily extended to inductively coupled nodes. To this end, we designate a secondary node to aid the communication between source and destination by acting as AF relay. We first consider the direct equivalent of the classical AF relaying, i.e. the node actively amplifies and retransmits the received signal using its own power source. This scheme, which we call active relaying, can be described within the circuit model introduced in the previous chapter. To avoid the problem of self-interference, we assume the relay operates in time division duplex (TDD) mode, i.e. the reception and retransmission take place in subsequent time slots. The circuit model describing the first time slot is shown in Fig. 4.1.

Here, the passive, frequency-dependent impedance matrix $\mathbf{Z}$ encapsulates all matching networks and the coupling impedance matrix of the node antennas. In the first time slot the source generates a signal with source voltage u_{S} and internal resistance R_{S}. This signal is received both by the relay and the destination in form of the voltages u_2 and u_3 over the impedances Z_{R} and R_{L}, respectively. These impedances hereby represent the input impedance of an LNA to further process the received signal at the both nodes. Both communication links can be described following the approach in Section 3.2.3, namely the effective SISO channel is obtained by terminating the ports of either the relay or the destination with the respective load impedance.

In the second time step, the relay transmits a scaled version of its previously received signal by acting as a source with voltage u_{R} and internal impedance Z_{R}. The source node is idle,

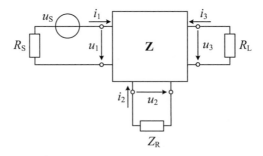

Figure 4.1. First step for inductively coupled active relaying.

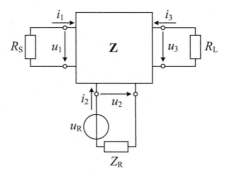

Figure 4.2. Second step for inductively coupled active relaying.

i.e. $u_S = 0$, and the input port of the source is terminated with R_S, therefore forming a SISO channel between relay and destination. Apart from the characteristics of the inductively coupled channel which differ from electromagnetic communication in the far-field, this scheme is identical to classical AF relaying as the voltage u_R represents an arbitrary scaled version—within the power constraints of the relay—of the signal received at the relay in the first time step.

4.2.2 Passive Relaying

In dense inductively coupled sensor networks, the mutual coupling of the nodes allows for an interesting alternative to active AF relaying. We refer to this alternative as passive relaying, as it allows for the implementation of the relay as purely passive circuit with no logic or power supply. To illustrate the concept, Fig. 4.3 shows the circuit model of a passive relaying setup. The source node again generates a signal with source voltage u_S. As indicated, the signal

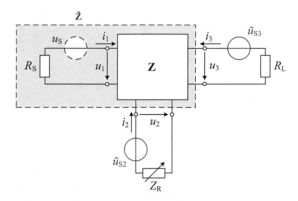

Figure 4.3. Circuit model for inductively coupled passive relaying. The dashed internal source is equivalently replaced by the open-circuit voltages $\tilde{u}_{S2}$ and $\tilde{u}_{S3}$.

source and the impedance matrix $\mathbf{Z}$ can be interpreted as an active two-port network. As discussed in Section 2.3, two-ports with internal sources can equivalently be represented as a passive two-port with voltage sources on both ports. In our setup, we denote the equivalent passive two-port as $\tilde{\mathbf{Z}}$, and the open-circuit voltages $\tilde{u}_{S2}$ and $\tilde{u}_{S3}$ are the effect of the source voltage at the relay and destination node, respectively. In this configuration the relay can be understood as a source with the voltage $\tilde{u}_{S2}$ and internal resistance Z_R. The total voltage $\tilde{u}_{S3}$ at the destination therefore contains signal contributions originating both from the source and the relay. Accordingly the relay load impedance Z_R affects the received signal. Similar to the optimization of gain coefficients in active relaying, the load impedance can therefore be optimized to improve communication performance. Furthermore, in contrast to the previously discussed active relaying scheme, passive relaying inherently operates in full-duplex mode.

Intuitively, the mechanism of passive relaying can be understood as the relay extracting signal power from the magnetic field generated by the source, which is in turn converted to a secondary field by the relay. To facilitate this effect, the passive relay is ideally implemented as a resonant circuit with a high quality factor. It is easy to show that network configurations exist in which the mere presence of a passive relay can then lead to a greater amount of power being extracted by the destination. We consider the network setup depicted in Fig. 4.4 as an example. Here the source and destination nodes are oriented orthogonally such that their

source   relay   destination

Figure 4.4. Example setup for passive relaying. The source and destination antennas are aligned such that $k_{SD} = 0$.

pairwise mutual coupling coefficient k_{SD} is zero. In consequence, the channel gain $|h|^2$ without the presence of the relay is also zero, as can be seen from (3.60). On the other hand the relay is arranged such that its pairwise couplings k_{SR} and k_{RD} to the source and destination are nonzero. The relay therefore extracts power from the field of the source and couples it to the destination.

To quantify the effect of the relay presence in this setting, we numerically evaluate the SNR for communication from source to destination. Specifically, the source is placed in the origin while the destination is located at $z = 30\,\text{cm}$ and rotated by an angle of $\alpha_D = 90°$. The relay is placed at $z = 15\,\text{cm}$ and oriented using all possible rotation angles $\alpha_R \in [0, 2\pi)$. All nodes use antennas with a radius of $r = 2.5\,\text{cm}$ and $\nu = 5$ windings. The relay has no matching network and its load impedance Z_L is a capacitor chosen to achieve resonance at the operating frequency of 25.4 MHz. The remaining simulation parameters are given in Tab. 3.1. Fig. 4.5 shows the SNR achieved for this setup as a function of the relay orientation. As

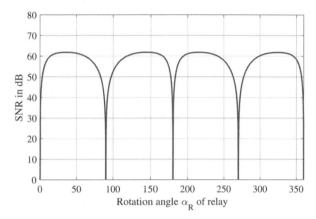

Figure 4.5. SNR of misaligned communication link using a passive relay.

can be expected, the SNR goes to zero at angles that produce either $k_{SR} = 0$ or $k_{RD} = 0$. Interestingly the optimum relay orientations are not found at angles of $\alpha_R = 45° + n90°$ for which, as follows from geometry, $k_{SR} = k_{RD}$. Instead the optima are shifted slightly closer to the orientation of the source, and found e.g. at 35.6°. The cause of this behavior is a detuning of the source, as will be discussed in Section 4.4.

Using passive relays to mitigate the impact of node misalignment is particularly valuable for the intended application of small-scale sensor networks in which the nodes often may have random and non-controllable orientations. In fact these random orientations lead SNR fluctuations which reduce the degree of interconnectivity in the network, similar to the fading caused in far-field communications by multipath propagation (cf. [161]).

4.3 Range Extension Using Passive Relays

The communication range of two nodes is inherently limited by the mutual coupling between them, as quantified in the coupling coefficient k. To analyze the impact of the coupling coefficient on the communication range, we restate the expression given in (3.60) for the maximum obtainable channel gain $|h|^2_{max}$ of an optimally matched, inductively coupled antenna pair operating at frequency ω:

$$|h|^2_{max} = \frac{\zeta}{\left(1 + \sqrt{1 + \zeta}\right)^2}. \tag{4.3}$$

We have herein introduced the shorthand notation

$$\zeta = k^2 \omega^2 \frac{L_m L_n}{R_m R_n}, \tag{4.4}$$

where the node antennas denoted by indices m and n exhibit self-inductances L_m and L_n and loss resistances R_m and R_n. The obtainable channel gain is asymptotically given by two different approximations depending on the value of ζ. For $\zeta \ll 1$, the channel gain may be approximated by

$$|h|^2_{max} \overset{\zeta \ll 1}{\approx} \frac{\zeta}{4}. \tag{4.5}$$

On the other hand, in the regime of $\zeta \gg 1$ we find that

$$|h|^2_{max} \overset{\zeta \gg 1}{\approx} 1. \tag{4.6}$$

The latter case has the interesting implication that even for coupling coefficients well below 1, the channel between the antenna pair can reach unit efficiency—given optimal matching—if the product of the intrinsic quality factors $Q_m = \omega L_m / R_m$ and $Q_n = \omega L_n / R_n$ is high enough to compensate the small value of k. A viable strategy for increasing the SNR of a source-destination pair separated by a large distance is therefore to use multiple relays placed between the pair such that neighboring relays are placed close enough together to be in the regime of high efficiency described by (4.6). This structure is also known in literature as magnetoinductive waveguide [134], as it enables the spatial transport of energy which is continually converted between a magnetic field (generated by the relay antennas) electric field (stored in the capacitances of the relay circuits).

To provide intuition on the conditions that this strategy can be employed in we investigate the case of a coaxially placed node pair separated by distance d, i.e. we choose $\mathbf{p}_m = [0, 0, 0]^T$ and $\mathbf{p}_n = [0, 0, d]^T$ in Cartesian coordinates, as well as $\mathbf{q}_m = \mathbf{q}_n = \mathbf{0}$. Without loss of generality we can assume $r_m \geq r_n$. For circular loop antennas in this configuration the

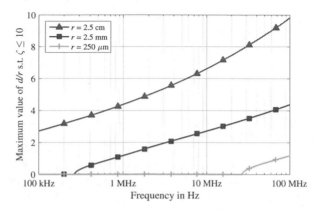

Figure 4.6. Visualization of maximum distance d/r for high efficiency regime of obtainable channel gain.

coupling coefficient can be approximated by [42]

$$k_{\text{coax}} \approx \frac{r_m^2 r_n^2}{\sqrt{r_m r_n}\left(\sqrt{d^2 + r_m^2}\right)^3}. \tag{4.7}$$

By substituting this expression into (4.4) and arbitrarily specifying $\zeta = 10$ as the lower limit for $\zeta \gg 1$ to be fulfilled, we obtain the condition

$$10 \leq r_m^3 r_n^3 \omega^2 \frac{L_m L_n}{R_m R_n} (d^2 + r_m^2)^{-3} \tag{4.8}$$

Solving for the distance, we find that the antenna pair is in the high efficiency regime for distances up to

$$d \leq \sqrt{r_m r_n \sqrt[3]{\frac{Q_m Q_n}{10}} - r_m^2}. \tag{4.9}$$

Fig. 4.6 visualizes the maximum distance at which (4.9) is fulfilled. Here both antennas have an identical radius r and $\nu = 5$ windings. We vary the operation frequency and normalize the maximum range to the antenna radius for better comparability among antenna sizes. The self-inductance and loss resistance were calculated from (3.58) and (3.27), respectively, assuming copper wire with radius $a = r/50$. It can be seen for the typically used frequencies in the kHz to MHz regime and using centimeter-scale antennas with a relatively low number of windings, the high efficiency regime can be utilized for the construction of magnetoinductive waveguides when the individual nodes are spaced apart up to a few radii. For smaller antenna sizes the intrinsic quality may need to be increased in order to enable utilization of the high-efficiency regime at typical node spacings. This may either be achieved by increasing the frequency or the number of windings.

4.3.1 Quasiperiodic Relay Configurations

The concept of magnetoinductive waveguides has been proposed by Shamonina et al. [134], [135]. To date the primary proposed applications for magnetoinductive waveguides are inductive power transfer [153], [2] and communication [31]. To the best of our knowledge, communication assisted by magnetoinductive relays has only been considered in quasiperiodic network arrangements [154], [156], [96]. A closed form capacity expression for 1D networks of equidistant nodes is presented in [157] assuming an additive white Gaussian noise (AWGN) model. The work also involves an analysis of the failure or misplacement of a single relay element. The capacity of this type of network has been optimized in [77] over the number of coil windings, carrier frequency, and transmit power spectral density (PSD), where the noise model was replaced by a circuit-based analysis. Transmit filter design and modulation schemes for digital communication over a 1D network is investigated in [76]. The authors of [22] analyze magnetoinductive communication over a 2D quasiperiodic grid of relays.

The focus of the literature on quasiperiodic structures of passive relays can be motivated by the substantial range extension observed for such configurations. As motivating example, we investigate the potential of passive relays to extend the usable communication range of a 1D coaxial node arrangement. A coaxially arranged source-destination pair with variable distance d, as already studied in Sec. 3.4, is used as reference setup. To extend the usable range of the communicating nodes, we insert N_S secondary nodes—acting as passive relays—into the reference network. These relay nodes are resonantly matched to the operating frequency of 25.4 MHz and coaxially positioned between the source-destination pair with equidistant spacings. Figs. 4.7 and 4.8 visualize the resulting SNR as a function of the distance d between the source-destination pair for the cases where the antenna radii of all nodes in the network have been chosen as $r = 2.5$ cm and $r = 250 \, \mu$m, respectively. All antennas have $\nu = 5$ windings, and the matching networks of the source and destination have been optimized for each network topology to maximize the SNR according to the conditions in (3.37) and (3.38).

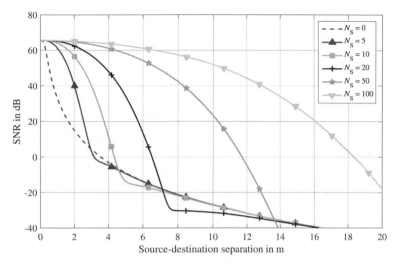

Figure 4.7. SNR over distance for 1D coaxial passive relaying with $r = 2.5\,\text{cm}$.

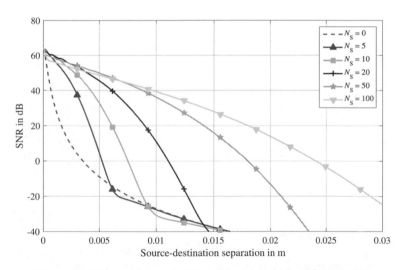

Figure 4.8. SNR over distance for 1D coaxial passive relaying with $r = 250\,\mu\text{m}$.

In both figures the dashed blue lines correspond to the reference case without relays, while the remaining curves represent cases with an increasing number of relays ranging from $N_S = 5$ to $N_S = 100$. The primary effect of the addition of relays can be observed as a reduction of the initial SNR decay rate over distance. This effect occurs in the regime where the relays experience sufficient coupling; when the relays become too far apart, the magnetoinductive wave guide effect breaks down and the SNR decay rate over distance approaches that of the case without relays. As figure of merit for the range extension achieved by the relays we again choose the distance at which the SNR drops to 0 dB, which we consider the usable range for communication. Hereby an interesting effect can be observed which leads to a counter-intuitive result for the case with $r = 2.5\,\text{cm}$: the scenario with $N_S = 5$ relays has a usable range of 3.2 m which is smaller than the value of 3.5 m for the relayless case. This effect can be attributed to the fact that choosing the relay load impedances $Z_{R,1}, \ldots, Z_{R,N_S}$ to achieve uncoupled resonance is not optimal with respect to the SNR for geometries with small node separations due to the detuning experienced for closely coupled resonant coils [21]. When further increasing the number of relays, however, the usable range is continually increased up to a range of 18.3 m when using $N_S = 100$ relays. For the setting with $r = 250\,\mu\text{m}$, the presence of passive relays increases the usable range from the reference value of 3.5 mm by up to 2.4 cm with $N_S = 100$ relays.

To further illustrate the detuning behavior, Figs. 4.9 and 4.10 visualize the channel power gain $|h(\omega)|^2$ achieved for different numbers of relays with $r = 2.5\,\text{cm}$. The channel gain is evaluated over frequency at fixed source-destination distances of $d = 3.5\,\text{m}$ and $d = 8\,\text{m}$, respectively. In both figures, the dashed blue curves again correspond to the reference network without relays, for which the source-destination pair is only weakly coupled in both cases. Accordingly the maximum of the reference channel power gain is found at the uncoupled design frequency of 25.4 MHz. For the case of $d = 3.5\,\text{m}$, it can be seen that the presence of relays leads to a slight detuning. The channel gain at resonance frequency is thereby reduced at this distance for $N_S = 5$. On the other hand, for $N_S = 20$ the relays experience sufficient coupling to exhibit a pronounced magnetoinductive waveguide effect which more than compensates the experienced shift in resonance frequency. In the second case with a source-destination distance of $d = 8\,\text{m}$, the results in Fig. 4.10 show that the network with $N_S = 20$ relay nodes experiences detuning which significantly impacts the channel gain at resonance frequency, while for $N_S = 5$ the relays are too weakly coupled to detune the system. The topic of mitigating the observed performance degradation due to detuning by optimizing the relay loads is addressed in the following section.

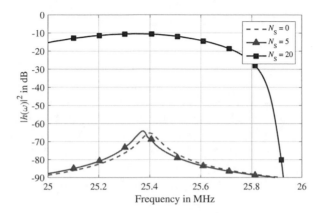

Figure 4.9. Channel power gain $|h(\omega)|^2$ at $d = 3.5\,\text{m}$.

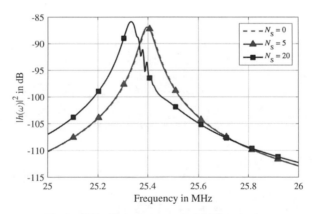

Figure 4.10. Channel power gain $|h(\omega)|^2$ at $d = 8\,\text{m}$.

4.4 Position and Load Optimization of Passive Relays

Most analyses of magnetoinductive structures arranged 1D or 2D structures consider only the coupling between adjacent relays. For a regularly spaced 1D coaxial setup, the authors of [82] investigate the impact of taking the coupling of nonadjacent nodes into account for a power transfer application, but find only slight deviations between the two coupling models. However, this insight can not generally be carried over to sensor networks and other ad hoc applications in which the wireless nodes are often randomly deployed, as there exist complex coupling relationships between all nodes in the network. In fact, we will show in Chapter 7 that considering all pairwise couplings in the network is essential for specific applications such as localization. One of the merits of the relaying channel model discussed in Sec. 3.2.3 is therefore the ability to analyze networks of arbitrary topologies.

When considering potential gains from passive relaying with no constraints on relay placement, it is sensible to question the optimality of the conventional wisdom found throughout literature of placing relays equidistantly in the previously investigated 1D coaxial setup. Additionally we have noted in the previous section that choosing the relay load impedance Z_R to achieve resonance at the operation frequency ω_0 under the assumption of no coupling is not optimal. We therefore provide intuition on optimal relay placement by considering a simple coaxial setup with a source-destination pair separated by distance d with the relay positioned at distance $d_R \in (0, d)$ from the source. Herein we also consider optimization of the relay load. Specifically we numerically optimize the SNR over the imaginary part of Z_R such that the source and destination matching networks are chosen to meet the conditions in (3.37) and (3.38) given each trial value of $\Im\{Z_R\}$. We leave the real part of Z_R—which represents the loss resistance of the relay antenna—constant, as increasing the real part simply reduces the quality factor of the relay, and reducing the real part is typically not possible without changing the physical antenna dimensions. The SNR achieved by the previously used relay loads achieving uncoupled resonance at the operating frequency $\omega_0 = 2\pi \cdot 25.4\,\mathrm{MHz}$ is used as reference.

Analysis of this relaying scenario suggests that the optimal relay position varies drastically for different coupling regimes. To illustrate this behavior Figs. 4.11 and 4.12 show the SNR as a function of the relay position, indicated by the normalized relay distance d_R/d, where the source-destination distance has been set to $d = 0.5\,\mathrm{m}$ and $d = 4\,\mathrm{m}$, respectively. In both curves the dashed black line indicates the SNR achieved when the relay is not present, while the blue and red curves correspond to the SNR using a relay with the default and optimized relay load, respectively. For the chosen antenna radius of $r = 2.5\,\mathrm{cm}$ and the distance of

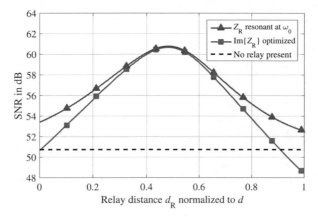

Figure 4.11. SNR for different relay placements with $d = 0.5\,\mathrm{m}$.

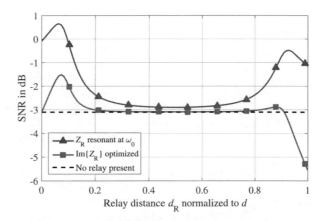

Figure 4.12. SNR for different relay placements with $d = 4\,\mathrm{m}$.

0.5 m, all nodes are strongly coupled. In this regime we observe the optimum relay position to be approximately at $d_R/d = 0.5$. Shifting the relay closer to source or destination leads to a reduced SNR. In addition the reference load of the relay, which achieves near-optimal SNR at $d_R/d = 0.5$, becomes increasingly suboptimal for relay positions close to the source or destination. This can be explained by the previously illustrated detuning experienced for closely coupled resonant coils. Optimizing the imaginary part of Z_R results in joint resonance at all three nodes, leading to significant SNR improvement for these positions compared to the reference load.

A different behavior is observed for the scenario with $d = 4$ m. For this setting the source-destination separation is greater than the previously determined usable range for the considered antenna size. As indicated by the results in Fig. 4.12 the achieved SNR has local maxima for a relay position close to the source or destination, with the minimum of the SNR for load optimization being at $d_R/d = 0.5$. This result is in contradiction to the conventional wisdom of placing the relay in the center [6].

The asymmetry of the curves in both scenarios can be explained by the definition the transmit power in the expression for the channel gain following as the squared magnitude of (3.45). Here the transmit power is considered as the active power flowing into the matching network of the transmitter. If the transmitter is detuned by the proximity of a relay, the resulting reduction in transmit power increases the channel gain by definition.

4.4.1 Random Relay Configurations

The results for the optimization of the one-dimensional relay position show that a passive relay does not necessarily need to be placed regularly between source and destination in order to achieve an SNR gain. This suggests that SNR gains are potentially also possible in a relaying configuration with random network topology. To investigate this conjecture we again consider the setting of a coaxial source-destination pair with a distance of $d = 0.5$ m. A single relay is randomly positioned according to a uniform distribution within a 3D bounding box of side length $b = 0.5$ m spanned by the source-destination pair, i.e. the relay has the position coordinates $x_R, y_R \sim \mathcal{U}(-b/2, b/2)$ and $z_R \sim \mathcal{U}(0, d)$. All nodes have identical orientations of $\mathbf{q} = \mathbf{0}$. Fig. 4.13 shows the empirical CDF of the resulting SNR with and without optimization of the relay load, while the dashed line corresponds to the reference SNR achieved without relay. The matching networks of source and destination were again optimized according to (3.37) and (3.38) for each realization.

For the case where the relay loads are not optimized, it can be seen that the presence of a

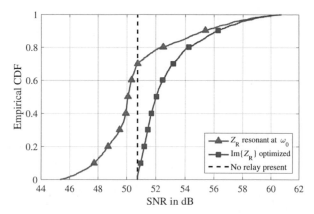

Figure 4.13. Empirical CDF of SNR for random relay position in 3D space. All nodes with $r = 2.5\,\mathrm{cm}$, $\nu = 5$, and identical orientation.

relay is not always beneficial. In fact approximately 70 % of the relay positions result in an SNR degradation of up to 5.4 dB with respect to the reference value obtained without relay, while for the remaining positions the SNR is increased by up to 9.9 dB. The performance loss can again be attributed to the detuning imposed by the relay. While optimizing the reactive part of the relay load impedance leads to a median SNR improvement of only 1.3 dB over the reference value, the optimization results in the relay improving the SNR for all observed realizations. We therefore expect that passive relaying is a viable mechanism to increase network interconnectivity in dense inductively coupled networks containing many nodes with the ability to act as relay, even if the nodes are not arranged in a regular structure.

4.4.2 Verification by Experiment

As proof of concept we experimentally measure the channel gain improvement by a passive relay in an inductively coupled communication link. We investigate a simple 1D coaxial setup with a source-destination pair separated by distance $d = 15\,\mathrm{cm}$ and a centrally positioned single relay with distance $d/2$ to both the source and the destination. The network is assembled and investigated using the localization measurement system described in Chapter 8. For source and destination, the node hardware and circuit layout are identical to the nodes used as localization anchors which are shown in Figs 8.2 and 8.1. The relay, on the other hand, is identical to a localization agent, which consists of a simple resonant circuit. The implemented node and its circuit are shown in Figs 8.4 and 8.3. All three nodes are jointly tuned to a resonance frequency of 22.5 MHz. The nodes were positioned using tripods as depicted in Fig 4.14.

We specifically measure the overall forward transmission gain $|S_{21}|^2$ between the ports of source and destination, both with and without the presence of the relays. This gain is proportional to the SNR according to the definition in (6.19) as the channel coefficient h is equivalent to S_{21} in this setup. It should be noted that the employed matching networks, which affect the measurement, are not optimized to maximize SNR. As seen from the results in Fig 4.15 the presence of the relay leads to a significant increase in SNR as expected. Here the dashed blue curve represents the reference case without the relay, showing a maximum value of $|S_{21}|^2 = -36\,\mathrm{dB}$. By adding the passive relay this value is increased to up to $-21\,\mathrm{dB}$, but the relay also introduces a slight detuning as indicated by the shift of the maximal gain to a higher frequency.

4.5 Conclusions

In large, dense microsensor networks, utilizing idle nodes as relays is a suitable method to overcome the limitations in communication range experienced in an inductively coupled physical layer due to misalignment and near-field decay rates. Using simulations and measurements, we verify a particularly favorable relaying approach for low-complexity microsensor networks, namely passive relaying, in which the relay nodes consist only of a passive, resonant circuit. From an investigation of joint position and load impedance optimization in simple coaxial arrangements we concluded that the conventional wisdom of equidistant relay placement does not apply in the loosely coupled regime. We furthermore demonstrate that the SNR gains of passive relaying can be achieved in the practically relevant setting of networks with random relay arrangements.

Figure 4.14. Setup for channel gain measurements using passive relaying.

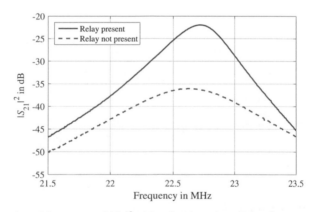

Figure 4.15. Measurement of $|S_{21}|^2$ with and without the presence of a passive relay.

79

5

Wireless Artificial Neural Networks

Beyond the mere extension of the communication range, which has been studied in the previous chapter, wireless relays have many interesting properties. In this chapter we present a novel approach to utilize relaying in a network of low complexity, inductively coupled sensor network with the goal of forming the wireless equivalent of an artificial neural network (ANN). We provide a method for programming the network functionality in a decentralized fashion and demonstrate the robustness of wireless ANNs against node failures and imperfections. Applications of this scheme exist in low-complexity sensor networks, where elaborate calculations can be carried out in a distributed fashion, or for creating powerful ANNs with very high degrees of interconnectivity realized by the wireless medium. Parts of this chapter are published in [142].

5.1   The Paradigm of Wireless Artificial Neural Networks

We envision inductively coupled wireless microsensor networks to typically comprise a large number of low-cost, low-complexity nodes. In most sensor networks with this design paradigm, the extraction and further processing of the data collected at the sensors is usually performed by a central unit with higher complexity, as the computational power of the wireless sensors is very limited. However, employing a central unit may negate some of the key benefits of sensor networks, such as ad-hoc deployment in almost any environment and low maintenance requirements.

Novel networking paradigms, such as node cooperation, applied with the goal of achieving elaborate computational abilities in a distributed manner, are therefore an open research topic. To this end, it has been recognized in literature that the interconnected topology of wireless sensor networks resembles the structure of artificial neural networks (ANNs), a model of computation abstracted from the interaction of biological neurons. Applying the principle of neurocomputation to wireless sensor networks is a natural choice, as the computation in ANNs is distributed, parallel, robust with respect to noisy data, and capable of learning. Consequently, recent research has proposed wireless implementations of artificial neural networks [39,78,109]. These approaches share the concept of implementing an ANN structure by reducing the broadcast nature of the wireless medium to point-to-point connections. In fact, as will be discussed in Sec. 5.2, the ability to control each connection between the nodes in the network separately is essential to achieving the desired computational properties. In state of the art wireless ANN approaches, this control is realized either by orthogonalization of signal transmission, e.g. in time or frequency, or by implementation of the ANN functionality in higher communication layers. Both mechanisms, however, have an undesirable scaling of used communication resources and complexity requirements of the nodes as the size of the network increases, rendering them infeasible for the considered setting of large-scale, low-complexity networks.

Instead we present an implementation of wireless ANNs that preserves the broadcast properties of the medium based on designating a subset of the nodes in the network as relays. Relay networks are known to improve communications e.g. by enabling orthogonalization of multiple data streams and have been well studied in literature (cf. [174], [16], and the references therein). By extending this approach, we use amplify-and-forward relaying to arbitrarily shape the channel matrix, which incorporates the channel as part of the desired computation. This allows for a narrowband and analog implementation of an ANN with low complexity constraints on the nodes even for large networks. We show that the configuration of the relays can be carried out in a distributed form, which further reduces overhead resulting from feedback. Furthermore, the robustness of the computation to error sources typically encountered in wireless networks is demonstrated.

5.2 System Model of Wireless Artificial Neural Networks

Artificial neural networks are mathematical models which mimic the computational mechanism encountered in biological neural networks. Stemming from this relationship, artificial neural networks are particularly efficient with tasks that can traditionally be solved better by humans than by computers, such as pattern recognition or object classification. The notion of artificial neural networks was first introduced by McCulloch et al. [98] in 1943, but many different models for ANNs have been developed since then. In this work, we will consider ANNs of the widely used multilayer perceptron form. For a broad introduction to ANNs and an overview of the various existing models, the reader is referred to [54] or [121].

In the following we develop a model for the wireless implementation of ANNs. We begin by providing a brief overview of the multilayer perceptron model.

5.2.1 The Multilayer Perceptron ANN

In multilayer perceptron ANNs several nodes are arranged in L sequential layers, as exemplified in Fig. 5.1. The first layer acts as input to the network and is accordingly called input

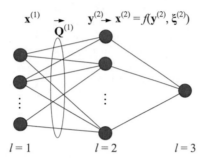

Figure 5.1. Multilayer feedforward ANN with $L = 3$ layers.

layer, while the last layer is referred to as output layer. The remaining layers are called hidden layers, as their operation is transparent to the overall functionality of the ANN.

The nodes in the input layer are referred to as input nodes. They generate—for example from sensed values—a data vector $\mathbf{x}^{(1)}$. This data vector is passed by the input layer further through the network using a series of interconnects. The nodes in the remaining layers are called neurons. As data is passed through the network in a time-discrete, feedforward fashion, the neurons and interconnects perform a manipulation of the input data. This process can be described as a concatenation of linear and nonlinear transformations. The linear part represents a weighting and superposition of transmitted data values, performed by the independent connections between each transmitting and receiving node. In general, the received data vector at the neurons of the $(l+1)$th layer is given by the linear relation

$$\mathbf{y}^{(l+1)} = \mathbf{Q}^{(l)}\mathbf{x}^{(l)}, \qquad l \in \{1, \ldots, L-1\}. \tag{5.1}$$

Here $\mathbf{x}^{(l)}$ represents the outputs of the neurons in the layer denoted by index l. These outputs are individually weighted at each connection to the inputs of subsequent layer. This weighting is modeled by the elements of the connection weight matrix $\mathbf{Q}^{(l)}$. Most discussions of neural networks in literature consider the elements of all quantities in (5.1) to be real-valued. We will adhere to this notion for the purposes of this analysis, although an extension to the complex domain is possible [57].

After receiving the signal vector $\mathbf{y}$, the neurons perform a nonlinear transformation described by their activation function $f(y, \xi)$. The basic and commonly employed notion of neuron functionality is to activate, i.e. choose a nonzero output value, if the respective input is greater than a threshold ξ. A simple way to implement this behavior can be achieved with the unit step function

$$f_{\text{unit}}(y, \xi) = \begin{cases} 0, & y < \xi, \\ 1, & y \geq \xi. \end{cases} \tag{5.2}$$

Another commonly used activation function is the logistic function[1], which is defined as

$$f_{\text{logistic}}(y, \xi) = \frac{1}{1 + \exp(y - \xi)}. \tag{5.3}$$

A particularly useful property of the logistic function is its continuous and easily calculated derivative, which facilitates the implementation of learning algorithms based on gradient descent methods.

The result of the transformation implemented by the activation function is passed as output to the subsequent layer. The output $x_m^{(l)}$ of the mth neuron in the lth layer is therefore defined as:

$$x_m^{(l)} = f\left(y_m^{(l)}, \xi_m^{(l)}\right), \qquad m \in \mathcal{M}^{(l)}, \qquad l \in \{2, \dots, L\}. \tag{5.4}$$

Here the input value y_m is the mth element of the receive vector $\mathbf{y}^{(l)}$, and $\mathcal{M}^{(l)}$ denotes an index set enumerating the neurons in the lth layer.

The result of the overall computation realized by the ANN is given by the output vector of the final layer, $\mathbf{x}^{(L)}$, which obviously depends on the weight matrices and activation functions of all layers. It is known in literature that multilayer perceptron ANNs are universal approximators [61], i.e. they can approximate any computation implemented by a continuous function with an arbitrary degree of precision, provided that the network is sufficiently large and the activation functions and weights are chosen appropriately. Universal approximation is possible even using feedforward ANNs with only a single hidden layer [27].

It should be noted that the weights implementing a specific computation may either (in

---

[1]The logistic function is also often referred to as sigmoid function due to its shape.

simple cases) be found by analysis, or, more practically relevant, by the use of learning algorithms. One example is the well-known backpropagation algorithm introduced in [127].

5.2.2 Wireless Implementation of ANNs on the Physical Layer

If we consider the interconnects between the neuron layers to be a wireless multiple-input, multiple-output (MIMO) channel, the linear input-output relation in (5.4) needs to be modified to describe the transformation of the data by the wireless channel. We assume that each layer transmits its output data in an analog broadcast, with the bandwidth and duration of the burst being below the coherence bandwidth and coherence time of the wireless channel, respectively. This leads to the following input-output relation:

$$\mathbf{y}^{(l+1)} = \mathbf{H}^{(l)}\mathbf{x}^{(l)} + \mathbf{n}^{(l+1)}, \qquad l \in \{1, \ldots, L-1\}, \tag{5.5}$$

For general wireless networks, the complex-valued entries of the channel matrix $\mathbf{H}^{(l)}$ describe a scaling and phase rotation of the transmitted signals. It should be noted that for wireless sensor networks with ad-hoc topology these values effectively appear random. Furthermore the additional term $\mathbf{n}^{(l)}$ is due to thermal noise at the receiving neurons. The network implements a specific computation if the condition

$$\mathbf{H}^{(l)} = \mathbf{Q}^{(l)}, \qquad \forall l \in \{1, \ldots, L-1\} \tag{5.6}$$

is fulfilled, i.e. if the channel matrix coincides with the desired weight matrix.

The key challenge of implementing a wireless ANN on the basis of a MIMO channel lies in the implementation of the desired weight matrices $\mathbf{Q}^{(l)}$ which may have arbitrary structure. The problem is illustrated in Fig. 5.2, which shows the degrees of freedom of weighting using wireless and wired interconnects between two layers. Specifically, we consider a set of neurons representing a subset of a larger ANN. The mth neuron in the lth layer generates the output symbol x_m. Assuming that the desired weight matrix $\mathbf{Q}^{(l)}$ is known, the contributions of x_m to the inputs of the neurons of the subsequent layer—denoted by y_{nm}, where $n \in \mathcal{M}^{(l+1)}$—need to be implemented by the interconnects in accordance with the mth column of $\mathbf{Q}^{(l)}$. For a wired infrastructure, as depicted in Fig. 5.2a, the value of x_m is transmitted over multiple independent wires. Each wired channel inherently scales the transmitted symbol with a complex scalar h_{nm} modeling the individual attenuation and phase shift of the wire. To implement the desired computation, each contribution can be scaled at the receiving neurons

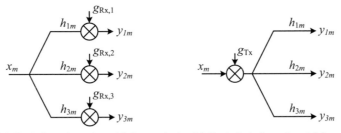

(a) Equivalent channel model for a wired ANN.

(b) Equivalent channel model for a wireless ANN.

Figure 5.2. Comparison between the weighting mechanisms for wired and wireless ANNs with a single transmitting neuron.

by a complex gain coefficient $g_{\text{Rx},n}$ which is simply chosen as

$$g_{\text{Rx},n} = \frac{[\mathbf{Q}^{(l)}]_{nm}}{h_{nm}}. \tag{5.7}$$

For wireless interconnects the contributions from the transmitted symbol x_m are likewise scaled by complex coefficients h_{nm} corresponding to the entries of $\mathbf{H}^{(l)}$. However, while wired ANNs can control each connection weight independently, the wireless receiving neurons can not differentiate between the individual contributions[2] which are received superimposed with the signals from all other transmitting neurons. All contributions pertaining to x_m can therefore only be scaled jointly at the mth transmitter, as depicted in Fig. 5.2b. This method does not offer the necessary degrees of freedom to realize arbitrary weighting operations $\mathbf{Q}^{(l)}$ over random channels $\mathbf{H}^{(l)}$ as the space of achievable weight matrices is limited to $\mathbf{Q}^{(l)}$ having the property

$$\mathbf{Q}^{(l)}[:, n] = c_n \mathbf{H}^{(l)}[:, n] \qquad \forall\, n, \tag{5.8}$$

where c_n are arbitrary constants and $\mathbf{Q}^{(l)}[:, n]$ denotes the nth column of $\mathbf{Q}^{(l)}$.

5.2.3 Channel Shaping by Cooperative Relaying

As previously discussed, a straightforward means of achieving wireless interconnects that resemble a desired weight matrix $\mathbf{Q}^{(l)}$ is to orthogonalize the transmission in time or frequency,

---

[2]We implicitly assume that each neuron only uses a single antenna.

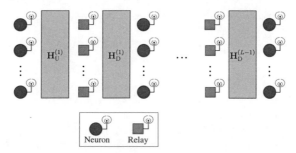

Figure 5.3. System model of wireless ANN with L neuron layers. The input data is passed through the network in a multihop fashion.

such that each orthogonal resource block is effectively a single-input single-output (SISO) transmission system where the desired weighting according to (5.7) can be performed at the transmitting or receiving neuron. The drawback of this scheme is that the resources required to transmit the output symbols of layer l to the neurons in layer $(l + 1)$ increase by a factor of $|\mathcal{M}^{(l)}| \cdot |\mathcal{M}^{(l+1)}|$, which becomes particularly prohibitive for large networks.

Instead we propose to implement appropriately weighted wireless interconnects as a MIMO channel by utilizing the concept of cooperative relaying. Following the notion of a dense wireless sensor network, we assume that there are several otherwise idle nodes available to assist the communication between the nodes in two layers of neurons by acting as relays. Analogous setups have been thoroughly studied in the context of wireless communications. For example, multiple schemes have been presented that employ a set of coherent, i.e. phase synchronized AF relays with transmit power constraint to improve a MIMO channel. By selecting the complex relay gains appropriately, it has been shown that a rank-deficient channel can be improved [173] or the effective channel matrix can be diagonalized [174]. In the following, we extend this setup with the goal of selecting relay gains which realize arbitrary effective channel matrices according to the requirements of the wireless ANN.

Our approach employs $N_{\mathrm{R}}^{(l)}$ nodes as coherent relays between the lth and $(l + 1)$th neuron layer, leading to the network structure given in Fig. 5.3. The relays apply a half-duplex amplify-and-forward operation, meaning the output of the relays is given by their respective input, scaled by a complex gain coefficient $g_i^{(l)}$, and with an additional noise term $m_i^{(l)}$. Here, i is the relay index: $i \in \{1, \ldots, N_{\mathrm{R}}\}$. Introducing the relays splits the original channel matrix $\mathbf{H}^{(l)}$ into an uplink channel from the lth neuron layer to the subsequent relays, denoted as $\mathbf{H}_{\mathrm{U}}^{(l)}$, and a downlink channel between the relays and the $(l + 1)$th neuron layer, $\mathbf{H}_{\mathrm{D}}^{(l)}$. With this model, the input-output relation given in (5.5) is replaced by

$$\mathbf{y}^{(l+1)} = \mathbf{H}_D^{(l)} \mathbf{G}^{(l)} \mathbf{H}_U^{(l)} \mathbf{x}_{(l)} + \mathbf{H}_D^{(l)} \mathbf{G}^{(l)} \mathbf{m}^{(l)} + \mathbf{n}^{(l+1)}, \tag{5.9}$$

where $l \in \{1, \ldots, L-1\}$, the matrix $\mathbf{G}^{(l)} = \text{diag}\{\mathbf{g}^{(l)}\}$ contains the gains of the lth relay layer, and $\mathbf{m}^{(l)}$ is the corresponding relay noise vector.

It should be noted that this presented channel shaping scheme holds for general wireless networks, both operating in the far-field as well as the inductively coupled near-field. The relays are then either classical AF relays or active inductively coupled relays, as discussed in Chapter 4. The only difference in a mathematical sense is that for the far-field channels the coefficients of $\mathbf{H}_D$ and $\mathbf{H}_U$ are generally complex valued, while for inductively coupled systems the channels are generally real-valued[3].

The desired computation is implemented with this network configuration if the choice of relay gains fulfills the following condition for all effective channels:

$$\mathbf{H}_D \mathbf{G} \mathbf{H}_U = \mathbf{Q}. \tag{5.10}$$

We have dropped the layer index l here for convenience of notation. The relay gains satisfying (5.10) can be found analytically by solving the linear problem

$$\tilde{\mathbf{H}} \mathbf{g} = \text{vec}\{\mathbf{Q}^T\}, \tag{5.11}$$

where the operator $\text{vec}\{\cdot\}$ represents the stacking of the columns of its argument into a vector. The matrix $\tilde{\mathbf{H}}$ being given by

$$\tilde{\mathbf{H}} = \left[\mathbf{1}_{N_{Rx}} \otimes \mathbf{H}_U^T \right] \odot \left[\mathbf{P}_{N_{Tx}} \left(\mathbf{1}_{N_{Tx}} \otimes \mathbf{H}_D \right) \right]. \tag{5.12}$$

Herein, $\mathbf{1}_n$ is the n-dimensional all-ones vector, $N_{Tx} = |\mathcal{M}^{(l)}|$ is the number of transmitting neurons and $N_{Rx} = |\mathcal{M}^{(l+1)}|$ the number of receiving neurons. The operators $\otimes$ and $\odot$ further denote the Kronecker and Hadamard product respectively, and $\mathbf{P}$ is a permutation matrix which performs a sorting of rows in the right-hand term. More intuitively, $\tilde{\mathbf{H}}$ can equivalently

---

[3]This is a consequence of the delay being small compared to the frequencies of interest for inductive coupling, such that it is neglected in the MQS approximation (cf. Chapter 3).

be written as

$$\tilde{\mathbf{H}} = \begin{bmatrix} \mathbf{H_U}^\mathrm{T} \\ \mathbf{H_U}^\mathrm{T} \\ \vdots \\ \mathbf{H_U}^\mathrm{T} \end{bmatrix} \odot \begin{bmatrix} \mathbf{H_D}[1,:] \\ \vdots \\ \mathbf{H_D}[1,:] \\ \mathbf{H_D}[2,:] \\ \vdots \\ \mathbf{H_D}[2,:] \\ \vdots \end{bmatrix} . \tag{5.13}$$

Equation (5.12) has a unique solution for each relay layer if the number of used relays is $N_\mathrm{R} = N_\mathrm{Tx} \cdot N_\mathrm{Rx}$. If N_R is larger than this required number, there exist multiple solutions. In this case it is suitable to select the gain vector $\hat{\mathbf{g}}$ with minimal length to minimize the sum transmit power of the relays. This solution is obtained using the Moore-Penrose pseudoinverse as

$$\hat{\mathbf{g}} = \tilde{\mathbf{H}}^+ \mathrm{vec}\left\{ \mathbf{Q}^\mathrm{T} \right\}, \tag{5.14}$$

where $(\cdot)^+$ denotes the pseudoinverse operation.

5.3 Wireless ANN Behavior in Additive Noise

As for all types of wireless communications, we can expect thermal noise to have a defining impact on the overall performance of the proposed wireless ANN scheme. To investigate this impact, we initially analyze the performance of the scheme using a generalized wireless network model with the channel coefficients being independent and identically distributed (i.i.d.) following a Gaussian distribution, i.e. $\mathbf{H_U}, \mathbf{H_D} \sim \mathcal{CN}(\mathbf{0}, \mathbf{I})$ with $\mathbf{I}$ denoting the identity matrix. The noise vectors are also assumed to be Gaussian, $\mathbf{m}, \mathbf{n} \sim \mathcal{CN}\left(0, \sigma_\mathrm{N}^2\right)$. In Section 5.5 we will extend the performance analysis to wireless ANN operating on channels formed by inductively coupled sensor nodes.

5.3.1 Comparison of Ideal and Wireless ANNs

We begin by investigating the overall feasibility of applying cooperative relaying to implement a wireless ANN. To this end we carry out a simulative analysis comparing an ideal, wired ANN to its proposed wireless counterpart. Specifically we consider an example ANN designed to perform character recognition on noisy data. The structure of this simulation is shown in Fig. 5.4. The input data to the ANN are the binary pixel values $x_m^\in \{-1, 1\}$ of 7 by 5 bitmaps

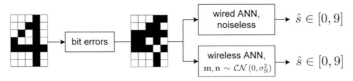

Figure 5.4. Comparison of wireless and wired implementations of an ANN performing character recognition.

representing the digits zero through nine. The noise in the input data is modeled by adding Gaussian distributed noise with zero mean and unit variance to the true values of the pixel data and taking the sign of the result, leading to a bit error probability of the input data of approximately 0.16.

Both the wired and the wireless networks have an identical neuron structure exhibiting an input layer with $7 \cdot 4 = 35$ input nodes, an output layer with 10 neurons each representing a specific character hypothesis, and no hidden layers. The weight matrix $\mathbf{Q}$ is chosen such that its rows represent the patterns of the respective hypothesis. The magnitude of the output generated by the network is then a measure of the closeness of the input to the individual character hypotheses. The estimate of the true digit represented by the bitmap data therefore corresponds to the index of the neuron producing the largest output value.

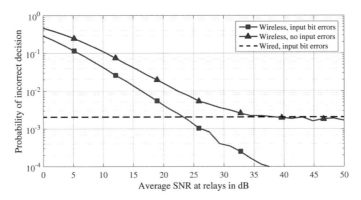

Figure 5.5. Error performance of wireless and wired character recognition ANNs.

In Fig. 5.5, the simulated probability of the wireless ANN choosing an incorrect digit is shown in blue as a function of the signal-to-noise ratio (SNR) due to the thermal noise at the relays and neurons. Specifically we consider the average SNR at the relay nodes, which simply equates to $1/\sigma_N^2$. For comparison, the black line shows the performance of the equivalent wired implementation of the same network, which is assumed to be noiseless. The observed error floor is due to the random bit errors in the input pattern. It can be seen that the error probability of the wireless case well matches that of the conventional implementation if the SNR reaches sufficiently high levels of about 35 dB. For the case of error free input patterns, the performance of the wireless network is given by the red curve, which shows an additional improvement in error performance and no saturation for high SNR.

5.3.2 Impact of Noise Amplification

To evaluate the behavior of wireless ANNs in more detail, it is helpful to study a simpler and more intuitive network than the previous example used for character recognition. A canonical and well understood example in the theory of neural networks is the XOR network, which implements a Boolean exclusive OR (XOR) operation on its input bits. A possible implementation of the XOR operation is given by a network with $L = 3$ layers, in which all neurons use identical activation functions chosen as a shifted unit step function with $\xi = 0.5$, i.e. $f(y, \xi) = f_{\text{unit}}(y - 0.5)$. There are two nodes each in the input and hidden layer, and a single node in the output layer, with the weight matrices chosen according to

$$\mathbf{Q}^{(1)} = \begin{bmatrix} 1 & -1 \\ -1 & 1 \end{bmatrix} \text{ and } \mathbf{Q}^{(2)} = \begin{bmatrix} 1 \\ 1 \end{bmatrix}. \tag{5.15}$$

The resulting ANN is shown in Fig. 5.6, from which it can easily be seen that for binary input vectors $\mathbf{x}^{(1)} \in \{0, 1\}^2$ the output $y^{(3)} \in \{0, 1\}$ gives the result of the XOR operation.

A noteworthy effect of the use of cooperative relaying is demonstrated in Fig. 5.7. Here the probability of incorrectly computing the XOR operation has been evaluated simulatively for 2000 random realizations of a wireless XOR network using the minimum number of relays, i.e. 4 relays between input and hidden layer, and 2 relays between hidden and output layer. We again consider a generalized wireless channel model with all channels being independently drawn from a zero-mean Gaussian distribution with variance σ^2. Specifically, we evaluate two cases for the channel variance, namely $\sigma^2 = 1$ and $\sigma^2 = 0.01$. These values can be interpreted as modeling environments with different degrees of inherent path loss. In the context of this analysis the relay gains in (5.14) are not subject to a power constraint, meaning that

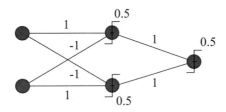

Figure 5.6. Artificial neural network implementation of the Boolean XOR operation.

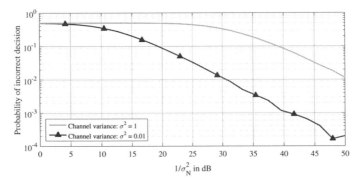

Figure 5.7. Probability of incorrect XOR decision for different channel attenuations (identical noise).

the magnitude of the relay gains can be arbitrarily large to offset the attenuation from the channel. For the same noise variance, however, the networks with channel variance $\sigma^2 = 0.01$ significantly underperform with respect to the networks with $\sigma^2 = 1$, corresponding to a loss of approximately 10 dB in SNR. The reason for the vastly different performance of both cases lies in noise amplification at the relays, as an increase in gain magnitude to compensate the path loss produces stronger contributions from the relay receive noise vector $\mathbf{m}$.

A straightforward solution to the problem of noise amplification is a scaling of the desired weights and thresholds. As an example, the XOR network in Fig. 5.6 can be modified such that $\tilde{\mathbf{Q}}^{(1)} = c\mathbf{Q}^{(1)}$, $\tilde{\mathbf{Q}}^{(2)} = c\mathbf{Q}^{(2)}$, and $\tilde{\xi} = c\xi$ with c being a real-valued constant. This scaling retains the computational functionality of the network, and the gain requirements of the relays can be reduced by choosing $c < 1$. Interestingly, in the context of inductively coupled networks a sufficient reduction in required gain magnitude can enable the implementation of wireless ANNs using the notion of passive relays introduced in Chapter 4.

Another approach to reduce the relay gains and the associated power consumption can be found by incorporating further nodes as relays. Keeping in mind that the minimal number of relays to find a unique solution for the relay gains in (5.12) is $N_{\text{Tx}} \cdot N_{\text{Rx}}$ for each layer respectively, this means that the number of relays is increased to $N_{\text{Tx}} \cdot N_{\text{Rx}} + N_{\text{ex}}$, where the number N_{ex} is called excess relays. As the typical system of interest for these considerations is a wireless sensor networks containing a large number of nodes, it can be assumed that sufficiently many additional nodes are available to serve as excess relays. As choosing the relay gains in the resulting underdetermined system according to (5.14) yields the gain vector with the minimal norm, the gains of the individual relays are expected to decrease, reducing the impact of noise amplification.

We demonstrate this behavior by repeating the simulation of random XOR networks with a channel variance of $\sigma^2 = 0.01$ for different amounts of excess relays. The resulting probabilities of incorrectly deciding on the XOR operation are shown in Fig. 5.8 as a function of the noise variance σ_{N}^2. Here, the dashed and dotted curves correspond the cases where 5 and 10 excess

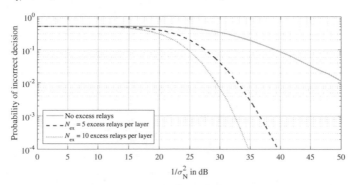

Figure 5.8. Propability of incorrect XOR decision for different excess relays. Channel variance $\sigma^2 = 0.01$.

relays have been introduced per layer, respectively. It can be seen that the error probability is significantly reduced as more excess relays are added to the network. To verify that the performance gains stem from reduced noise amplification, we evaluate the realization-averaged power gain of the relays in the first layer, $\mathsf{E}\left[\frac{1}{N_{\text{Tx}} \cdot N_{\text{Rx}} + N_{\text{ex}}} \left\| \mathbf{g}^{(1)} \right\|^2\right]$. The average gain reduces from 43.6 dB for networks without excess relays to 31.4 dB for $N_{\text{ex}} = 5$ and 25.7 dB for $N_{\text{ex}} = 10$, which matches the expected behavior.

5.4 Iterative Gain Allocation with Reduced Feedback

Finding the appropriate relay gains to implement a computation using (5.14) requires full knowledge of $\mathbf{H}_\mathrm{U}^{(l)}$ and $\mathbf{H}_\mathrm{D}^{(l)}$. The overhead generated by disseminating this channel state information (CSI) scales with the number of relays and neurons. To alleviate this potential drawback, we propose an iterative and decentralized method for the relays to individually choose their gains based solely on their local CSI and feedback independent of the number of relays. Due to the decentralization, this approach may prove particularly beneficial in large networks.

To formulate the approach, we express the gain allocation for each layer as the minimization problem

$$\hat{\mathbf{g}} = \arg\min_{\mathbf{g}} \|\mathbf{H}_\mathrm{D}\mathbf{G}\mathbf{H}_\mathrm{U} - \mathbf{Q}\|_\mathrm{F}^2 , \tag{5.16}$$

where $\|\cdot\|_\mathrm{F}$ denotes the Frobenius norm. A solution to this optimization problem can be found numerically. To this end we apply a heuristically modified gradient decent algorithm originally presented in [123], where the scheme has been introduced to perform decentralized relay gain allocation for sum-rate maximization in a relay-assisted, two-hop MIMO communication setup.

The decentralized properties of the optimization can be seen from the calculation of the complex gradient of the norm in (5.16), which consists of the elements

$$\frac{\partial \|\mathbf{H}_\mathrm{D}\mathbf{G}\mathbf{H}_\mathrm{U} - \mathbf{Q}\|_\mathrm{F}^2}{\partial g_k^*} =$$
$$\mathrm{tr}\left\{\mathbf{H}_\mathrm{D}\mathbf{G}\mathbf{H}_\mathrm{U} \cdot \mathbf{H}_\mathrm{U}^\mathrm{H}\mathbf{E}_n^\mathrm{H}\mathbf{H}_\mathrm{D}^\mathrm{H} - \mathbf{Q} \cdot \mathbf{H}_\mathrm{U}^\mathrm{H}\mathbf{E}_n^\mathrm{H}\mathbf{H}_\mathrm{D}^\mathrm{H}\right\} \tag{5.17}$$

Here, $(\cdot)^*$ denotes the complex conjugate, $(\cdot)^\mathrm{H}$ is the Hermitian transposition, and $\mathbf{E}_n$ is a matrix with all zeros except for a one on the nth diagonal position. The desired weight matrix $\mathbf{Q}$ can be assumed to be known a priori. Furthermore, the term $\mathbf{H}_\mathrm{U}^\mathrm{H}\mathbf{E}_n^\mathrm{H}\mathbf{H}_\mathrm{D}^\mathrm{H}$ corresponds to the local CSI at the nth relay, and the term $\mathbf{H}_\mathrm{D}\mathbf{G}\mathbf{H}_\mathrm{U}$ is the effective channel between the neuron layers, which can be fed back in a broadcast fashion by the receiving neurons. As the dimensions of these terms do not scale with the number of relays, the overhead required for CSI dissemination grows slower with N_ex than for methods requiring global CSI, such as the pseudoinverse based approach, making the gradient search well suited for large networks containing many relays. However, the optimization cost function in (5.16) does not consider the amplification of noise received by the relays, which may have an impact on the error performance of the overall computation. We will therefore call the previously introduced scheme *gradient A*, and also investigate an alternative problem formulation denoted *gradient B*, which instead attempts to minimize the mean squared error in the signal received by the neurons. The new cost function is then

$$\hat{\mathbf{g}} = \arg\min_{\mathbf{g}} \mathsf{E}\Big[\, \|(\mathbf{H}_\mathrm{D}\mathbf{G}\mathbf{H}_\mathrm{U} - \mathbf{Q})\mathbf{x} + \mathbf{H}_\mathrm{D}\mathbf{G}\mathbf{m} + \mathbf{n}\|_\mathrm{F}^2 \,\Big], \tag{5.18}$$

where we use the operator $\mathsf{E}[\cdot]$ to denote expectation with respect to $\mathbf{x}$, $\mathbf{m}$, and $\mathbf{n}$. We can again calculate the complex gradient, which now has the elements

$$\frac{\partial \mathsf{E}\Big[\, \|(\mathbf{H}_\mathrm{D}\mathbf{G}\mathbf{H}_\mathrm{U} - \mathbf{Q})\mathbf{x} + \mathbf{H}_\mathrm{D}\mathbf{G}\mathbf{m} + \mathbf{n}\|_\mathrm{F}^2 \,\Big]}{\partial g_k{}^*} =$$
$$\mathrm{tr}\big\{\mathbf{H}_\mathrm{D}\mathbf{G}\mathbf{H}_\mathrm{U} \cdot \boldsymbol{\Sigma}_\mathbf{x} \cdot \mathbf{H}_\mathrm{U}^\mathrm{H}\mathbf{E}_n^\mathrm{H}\mathbf{H}_\mathrm{D}^\mathrm{H} - \mathbf{Q}\boldsymbol{\Sigma}_\mathbf{x} \cdot \mathbf{H}_\mathrm{U}^\mathrm{H}\mathbf{E}_n^\mathrm{H}\mathbf{H}_\mathrm{D}^\mathrm{H}$$
$$+\, \sigma_\mathrm{N}^2\mathbf{H}_\mathrm{D}\mathbf{G}\mathbf{E}_n^\mathrm{H}\mathbf{H}_\mathrm{D}^\mathrm{H}\big\}\,. \tag{5.19}$$

This gradient is similar to (5.17), but now considers the covariance matrix of the input signal, $\boldsymbol{\Sigma}_\mathbf{x}$, and the noise contribution term $\sigma_\mathrm{N}^2\mathbf{H}_\mathrm{D}\mathbf{G}\mathbf{E}_n^\mathrm{H}\mathbf{H}_\mathrm{D}^\mathrm{H}$.

To evaluate the suitability of the two presented decentralized gain allocation schemes, we apply them to find the relay gains achieving the Boolean XOR operation as specified by the weight matrices in (5.15).

The key parameters used in the analysis of this network are the number of excess relays N_ex and the noise variance σ_N^2, which have been chosen identically for all layers. We simulate the error probability of the XOR operation, again using i.i.d. Gaussian channels with zero mean and unit variance, as well as a noise variance of $\sigma_\mathrm{N}^2 = 10^{-3}$. Figure 5.9 shows the error as a function of the SNR for N_ex equal to 0 (shown in blue), 5 (green), and 20 (red) relays. The dashed lines represent relay gain allocation using the closed form minimum norm approach in (5.14), whereas the square and triangle markers represent gradient schemes A and B, respectively. It can be seen that the error probability of the minimum norm scheme improves with the number of excess relays. For $N_\mathrm{ex} = 0$, the performance of gradient scheme A is equivalent to the closed form allocation, but as the number of excess relays is increased, it suffers from converging to suboptimal solutions. Gradient scheme B, however, does not converge to a sensible solution for the minimum relay configuration, but closely matches the performance of the minimum norm approach if the number of excess relays is increased.

We furthermore investigate the convergence behavior of both gradient schemes. Depicted

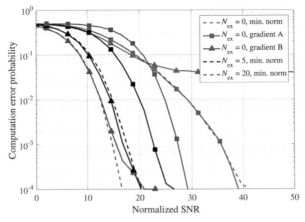

Figure 5.9. Error probability of the XOR network versus SNR for different relay gain allocation schemes.

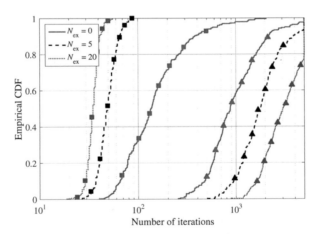

Figure 5.10. Convergence behavior of gradient based relay gain allocation schemes. (Square markers) gradient scheme A. (Triangle markers) gradient scheme B.

in Fig. 5.10 is the empirical CDF of the number of iterations the schemes perform before their solution meets the defined termination criteria and terminates.

The maximum number of permissible gradient search iterations, after which the schemes are forced to terminate, was set to 5000 for this analysis. Clearly, gradient scheme B (triangle markers) takes comparatively long to converge, and while it gains from an increased number of excess relays in terms of error probability, increasing N_{ex} negatively impacts its convergence speed. On the other hand, gradient scheme A (square markers) is not only faster in convergence, but even gains from an increased number of relays. Both schemes therefore have their benefits and drawbacks, and a choice between them would depend on the requirements of the implemented system.

5.5 Application of Wireless ANNs to Inductively Coupled Sensor Networks

Until now we have evaluated the concept of wireless ANNs using a simple and abstract network model with i.i.d. Gaussian channel coefficients. In this section, we will apply the concept of wireless ANNs to an inductively coupled sensor network and investigate the sensitivity of the computation to the additional imperfections that can be expected in such a practical sensor network environment. Specifically we consider the channel shaping to be carried out using inductively coupled, active AF relays as discussed in Section 4.2. To this end, each sublayer comprised of the transmissions from neurons to relays or relays to neurons was simulated separately, with the involved nodes being randomly placed within a 3D a box of dimensions $b \times b \times b$ and with identical orientations, i.e. $\mathbf{p}_n = [x_n, y_n, z_n, 0, 0, 0]^{\mathrm{T}}$ with $x_n, y_n, z_n \sim \mathcal{U}(-b/2, b/2)$. The roles of neurons and relays were randomly assigned within a given network. The channel matrices $\mathbf{H}_{\mathrm{U}}^{(l)}$ and $\mathbf{H}_{\mathrm{D}}^{(l)}$ as well as the noise variance σ_{N}^2 were evaluated using the circuit-based channel model presented in Section 3.2, where we assume node antennas with a radius of $r = 2.5\,\mathrm{cm}$ and $\nu = 5$ windings, as well as a transmit power of $P_{\mathrm{T}} = 1\,\mathrm{nW}$. For this setup, the SNR of the individual links is subject both path loss as well as the fading behavior resulting from node misalignment, as demonstrated in Chapter 4. To contextualize the inductively coupled setup with the previous results, Figure 5.11 therefore shows the empirical CDF of the resulting SNR where the side length b of the volume was varied from $b = 0.2\,\mathrm{m}$ to $b = 0.5\,\mathrm{m}$.

In consequence of the fading, it is possible that several of the wireless links of the ANN experience an outage situation: if a channel to or from a relay node is strongly attenuated as

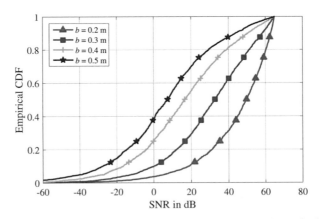

Figure 5.11. Empirical CDF of SNR in single transmission step with nodes confined to different volumes.

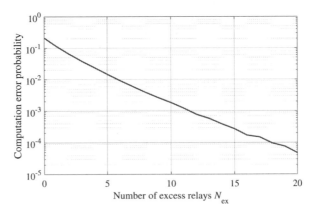

Figure 5.12. Error probability of XOR wireless ANN in inductively coupled network vs number of excess relays N_{ex}.

a result of the network geometry, the relay will compensate for the path loss by choosing a large gain $g_i^{(l)}$. However, as discussed in Section 5.3.2 this leads to strong noise amplification on the respective link. A possible mitigation strategy is to increase the number of excess relays N_{ex}. The randomly positioned relays introduce spatial diversity into the channels $\mathbf{H}_{\mathrm{U}}^{(l)}$ and $\mathbf{H}_{\mathrm{D}}^{(l)}$, which is exploited by the minimum weight relay gain assignment in (5.14). We demonstrate this behavior by simulating the computation error of the XOR network for the inductively coupled channels using a variable number of excess relays. Fig. 5.12 shows the error given a bounding box size of $b = 0.3\,\mathrm{m}$ for each sublayer. The results indicate that the fading is severe enough to render the XOR computation highly unreliable in the minimum relay configuration, i.e. for $N_{\mathrm{ex}} = 0$.

5.5.1 Node Failure

Sensor networks are known to be prone to node failures due to the low level of maintenance and low complexity of the devices. It is therefore thinkable that one of the nodes comprising a wireless ANN may cease to function. In case of a relay failure, the network may restore its computational functionality by reacquiring an updated choice of gains calculated over the remaining subset of relays, given the remaining number of relays still satisfies the minimum relay configuration. If on the other hand a neuron fails, the network will also need to define a new set of weight matrices, e.g. by learning. However, the parallel structure of ANNs suggests that computation might be robust to random node failures even without reconfiguring the network, and that this robustness can be increased by adding additional neurons and relays to perform the computation, thus adding redundancy. A particularly interesting case is that of a neuron ceasing to function. Using the XOR network, allow each node in the hidden layer[4] to fail on a random and independent basis with probability p_{N}. A neuron failure is implemented by setting its corresponding output value to $x_i^{(l)} \equiv 0$ irrespective of the input. The solid blue curve in Fig. 5.13 shows the computation error probability of the XOR network versus p_{N}, where we use a configuration with $N_{\mathrm{ex}} = 10$ excess relays per layer. For very low values of p_{N} below 10^{-3} the error probability is dominated by the thermal noise, and approaches 0.5 as p_{N} grows to 1, with a $1\,\%$ error probability being reached at $p_{\mathrm{N}} = 1.7 \cdot 10^{-2}$.

A simple strategy to increase the ANNs robustness against neuron failure is to extend the XOR network to a redundant version, consisting of multiple parallel XOR computation and a final stage of majority decision. This can be achieved within the framework of $L = 3$ layers by

---

[4]We restrict the random failures to the hidden layer as sensible computation is not possible if the output neuron fails.

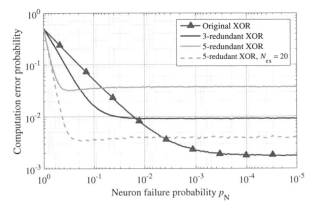

Figure 5.13. Error probability of XOR wireless ANN in inductively coupled network vs probability of neuron failure.

increasing the number of hidden nodes from 2 to $2n$, where n is an odd integer, and choosing the weight matrices

$$\mathbf{Q}^{(1)} = \mathbf{1}_n \otimes \begin{bmatrix} 1 & -1 \\ -1 & 1 \end{bmatrix} \text{ and } \mathbf{Q}^{(2)} = \mathbf{1}_{2n}. \tag{5.20}$$

The solid red and yellow curves in Fig. 5.13 show the error probabilities of the redundant XOR network for $n = 3$ and $n = 5$, respectively, with N_{ex} still set to 10. While increasing n leads to the error dropping off with a steeper slope for decreasing values of $\mathbf{p}_{\text{N}}$, it also leads to a saturation at higher error levels. This can be explained by the fact that the fraction of excess relays is smaller compared to the minimum requirement as n is increased, leaving less degrees of freedom for the channel shaping and thus a greater noise amplification by the relays. The dashed yellow curve therefore again shows the performance of the computation for $n = 5$, but with a setting of $N_{\text{ex}} = 20$. In effect, the error performance goes into saturation at a lower level as expected, and a 1 % computation error is possible for neuron failure probabilities up to $p_{\text{N}} = 4.3 \cdot 10^{-1}$.

5.5.2 Phase Synchronization Errors

ANNs implemented on low-complexity wireless nodes may exhibit error sources beyond receiver noise that are typically not encountered in their wired counterparts. One such issue is

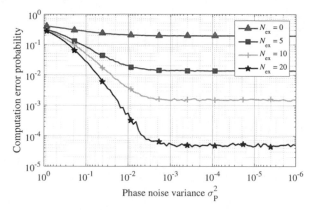

Figure 5.14. Error probability of XOR wireless ANN versus phase noise at relays. Relay weights selected using closed form solution.

the fact that the coherent relaying approach used in this work assumes perfect phase synchronization of the relays in each stage. Especially considering the low-complexity setting, the performance of a real network may suffer from synchronization inaccuracies. To investigate the impact of this, we model the imperfect synchronization as random phase offset for each relay gain drawn from a zero mean Gaussian distribution with variance σ_P^2, i.e. we choose the gains $\tilde{g}_i^{(l)} = g_i^{(l)} \cdot e^{jn_P}$ with $n_P \sim \mathcal{N}(0, \sigma_P^2)$. Figure 5.14 shows the computation error probability of the previously introduced XOR network as a function of the phase noise variance σ_P^2 where we investigate settings with different numbers of excess relays per layer. Here, the relay weights have been chosen using the closed form solution given in (5.14), and we again set $b = 0.3\,\mathrm{m}$. The behavior of the computation error in this setting is the result of two effects. Clearly the computation performance degenerates with a high level phase noise present in the system, as the channel shaping mechanism breaks down. This breakdown shows notable effects for phase noise variances in excess of $\sigma_P^2 = 3 \cdot 10^{-3}$. On the other hand, the computation error probability in the regime of low phase noise is defined by the number of excess relays and asymptotically approaches the performance without phase noise, as evaluated for the same setup in Fig. 5.12. It can be seen that increasing the number of excess relays can improve the error probability enough to still achieve acceptable error probabilities even for considerable phase noise variance.

5.6 Conclusions

We introduce the notion of wireless artificial neural networks implemented directly in the physical layer of multihop networks comprised of low-complexity nodes. By utilizing idle nodes to act as active relays we show that the channel matrix between the individual hops can be arbitrarily shaped if the number of relay nodes is sufficiently large. Using this channel shaping, wireless ANNs can be implemented in a broadcast fashion. This property presents an advantage of the proposed wireless ANN structure compared to ANNs with wired interconnects: while the wireless implementation allows for ANNs with arbitrarily many interconnects per hop without increasing the number of utilized orthogonal resources, the number of interconnects in wired ANNs is ultimately limited by space constraints, e.g. limitations in chip area [159].

The functionality of the ANNs investigated in this chapter is restricted to computations having a structure that can easily be derived. The true potential of ANNs, however, lies in the ability to adaptively conform to any desired relation between inputs and outputs by employing learning algorithms such as the well-known backpropagation algorithm [127]. To employ this algorithm within the framework of the proposed wireless ANN implementation, it needs to be modified to apply to the relay-assisted structure of network. This extension was presented in an undergraduate research project [111] along with a proof-of-concept application in which a wireless ANN is trained to perform RSS-based localization.

Part III.

Localization in Inductively Coupled Networks

6

Circuit Based Localization Using Multiple Anchors

In this chapter, we present a novel near-field localization method for inductively coupled networks which we refer to as circuit based localization. The approach is based on modeling the interacting nodes as two-port or multiport circuits. The unknown sensor position is hereby reconstructed from observations of network parameters such as the port input impedance or reflection coefficient. Circuit based localization reuses the communication hardware of the inductively coupled nodes and therefore requires no additional infrastructure besides the anchors and agents, although it can be improved by the presence of secondary nodes, as studied in Chapter 7. The scheme furthermore does not require the use of orthogonally arranged multi-axial antennas, and allows for the agent being implemented with purely passive circuit elements. In the combination of these defining aspects, circuit based localization is distinct from existing near-field localization methods as discussed in Section 6.1. It is therefore suitable for being employed in inductively coupled networks with many low-complexity inductively coupled devices, such as in-body sensor networks or RFID-type applications.

After providing an overview of existing localization methods for both far-field and near-field networks, we give a formulation of the circuit based localization approach and discuss the use of a priori knowledge of the cost function structure to decrease the numerical complexity of the localization process. We show by analysis of the theoretical performance bound of this approach as well as by simulative investigation that localization can be achieved by the proposed method with a high degree of accuracy. Parts of this chapter have been published in [139].

6.1   Wireless Localization Systems

We generally define localization in a wireless network as the process of obtaining estimates of the unknown spatial parameters of one or several nodes referred to as agents. The primary spatial parameter of interest is the node position, but several systems, including inductively coupled networks that are the focus of this work, also allow and in part require estimating the node orientation. Localization is performed on the basis of observations of signals or physical quantities within the network. In order to obtain estimates within a reference coordinate system, it is necessary that one or multiple nodes have a priori knowledge of their position and/or orientation. These nodes, referred to as anchors, are often also the point in the network at which the observations are obtained.

Having localization capabilities in a wireless network is highly desirable for many applications. The location information is often useful in its own right, for example in logistics where the wireless tracking of goods and vehicles is widespread [120]. Often the physical layer of the wireless communication in a network can be reused for localization, for example by using a network of wearable sensors to perform human motion tracking [100]. In the context of small scale sensor networks, sensor data can be associated with node positions, thus enabling a multitude of uses from spatial signal reconstruction [164] to event detection [86]. Furthermore, knowledge of the node location can be used to improve the communication within a wireless network. For example, the problem of routing in large ad hoc networks can be facilitated by taking the positions of the involved nodes into account. An overview of position-based routing protocols is provided in [97] and [155].

Wireless localization methods can be categorized into geometric and fingerprinting-based approaches. In the following, we will provide an overview of both categories and discuss prominent methods primarily in the context of far-field wireless systems. We furthermore discuss how localization methods differ for application in near-field systems. For further details on commonly used localization methods, the reader is referred to [93].

6.1.1   Geometric Localization Methods

Geometric localization methods reconstruct the position of an agent node by combining measurements with prior knowledge about the geometry of the network. Typically this knowledge refers to the positions of the anchor nodes. Geometric localization techniques are often a two-step process, in which metrics are extracted from exchanged signals in a first step. The localization is then based on these metrics afterwards. Three of the most prominently used signal metrics for localization are time-of-arrival (TOA), angle-of-arrival (AOA), and received signal strength (RSS).

By measuring the propagation time of signals between an anchor and multiple agents, TOA based localization systems obtain estimates of the pairwise distances between the nodes. This information is subsequently used to solve a trilateration problem to estimate the agent position. A well-known example of a practical systems using TOA measurements are satellite navigation systems such as GPS. AOA based localization systems, on the other hand, work by determining the relative angles between agent and anchor nodes. These angular estimates then form a triangulation problem to obtain the agent position. Finally, RSS based localization is based on the attenuation of a known transmitted signal due to relative position and orientation between anchor and agent. This class of localization schemes therefore requires an underlying path loss model. As an example, the received power P_r of a far-field system operating in free space can be modeled using the well-known Friis formula [44]

$$P_r = P_t G_t G_r \left(\frac{\lambda}{4\pi d} \right)^2 , \tag{6.1}$$

where P_t is the power of the transmitted signal, G_t and G_r are the respective antenna gain values (cf. [12]) of the transmitting and receiving antennas, λ is the wavelength of the signal and d is the distance between the devices. If the antenna gains can be assumed to be approximately omnidirectional, the pairwise distances between anchors and agent can be inferred directly from the received power of exchanged signals, resulting again in a trilateration problem to estimate the agent position.

6.1.2 Fingerprinting Based Localization Methods

In contrast to geometric localization, fingerprinting based localization methods require no prior knowledge of the network geometry. Instead, the process of localization is divided in two phases. During an initial training phase a metric of interest is observed for several known positions of the agent, and stored in a database together with the positions. As the process of extracting a metric from a received signal potentially reduces the information available about the agent's position [136], it is also possible to store the sampled signal itself. In the subsequent localization phase, the fingerprinting localization system acquires a measurement of the previously recorded quantity with the agent being at an unknown position. Using a similarity measure between the measured and prerecorded data, the agent position can be estimated. A detailed discussion of fingerprinting based localization can be found in [151].

6.1.3   Localization in Near-Field Systems

A fundamental difference between far-field and near-field localization approaches lies in the rate of decay of the information carrying field quantity. It can be shown that in the far-field region, i.e. at distances normalized to the wavelength λ such that $\frac{2\pi}{\lambda}d \gg 1$, the amplitudes of both the electric and magnetic field decay with increasing distance d to the source as $\mathbf{E} \propto 1/d$ and $\mathbf{B} \propto 1/d$. In the near-field region ($\frac{2\pi}{\lambda}d \ll 1$), however, the rates of decay are $\mathbf{E} \propto 1/d^2$ and $\mathbf{B} \propto 1/d^3$ (cf. [12]). While the higher rate of decay limits the useful range of interaction between devices, the steeper spatial gradient of the field strength is in fact beneficial for localization methods based on RSS, yielding more precise position estimates in the near-field.

In addition, in the specific context of inductively coupled sensor networks, some of the above mentioned far-field localization approaches are suboptimal. In order to increase the efficiency of energy transfer between the nodes, inductively coupled systems are often designed resonantly, leading to a significantly smaller signaling bandwidth than used in comparable far-field communication systems. The ranging process in TOA based localization suffers from such narrow bandwidth, as the temporal resolution of exchanged signals decreases [166]. On the other hand, AOA based localization would require employing antenna arrays at least on a subset of the nodes in the network, thus adding otherwise unwanted complexity. The concept of RSS based localization is applicable to inductively coupled sensor networks in principle, but can not be based on propagating waves as wave propagation is negligible at the frequencies of interest.

It is therefore sensible to base the localization in near-field systems on the observation of quantities related to the magnetic near-field interaction. For localization in inductively coupled systems many existing schemes reconstruct the unknown position of an agent node using measurements of voltages induced by the magnetic field of a driven source node at one or several points in space. As the magnetic flux density $\mathbf{B}$ is a vector quantity, these approaches often employ antenna structures with two or three orthogonal loop antennas for the source or the measuring nodes. For example, Raab et al. [118] have proposed a method for magnetic localization using a single tri-axial field source, which acts as anchor, and single tri-axial sensor acting as agent. Based on a vector measurement of the induced voltages at the agent for a known vector of excitation currents at the anchor, the system is able to reconstruct the unknown position and orientation of the agent using a system model based on the dipole approximation of mutual inductance (cf. Section 3.1.2). A variation of this idea is presented in [112], where a rotating field is generated by a two-axis anchor. This field is observed at the agent coils over time to determine their position.

It is also possible to reconstruct the agent position if both source and measurement devices employ single-axis loop antennas. In this case it is necessary to employ multiple anchor nodes acting as source or measurement device, to resolve the positional and orientational ambiguity of the agent. As an example, an in-body localization system has been demonstrated in [84], which operates by measuring the voltages induced at a capsule by several anchors acting as sources, and transmitting these measurements to an external workstation for further processing. A similar approach is used in [95], where underground localization of badgers is performed by measuring anchor-induced voltages that are collected and transmitted to a central unit using an IEEE 802.15.4 link when the badger is at the surface.

Localization schemes for inductively coupled devices may also employ additional infrastructure to facilitate the localization process. For example it is possible to measure the magnetic field generated by a driven agent using specialized measurement instruments which are not part of the sensor network. Many localization systems are described in literature—especially in the medical field—which locate a magnetic source using magnetometers[1]. Localization schemes with very good accuracy have been presented employing e.g. superconducting quantum interference devices (SQUIDs) [169, 170] or fluxgate magnetometers [110]. Particularly SQUIDs are able to detect very small fluctuations of magnetic fields in the sub-nT range. Even though SQUID sensors are small enough to be manufactured on a chip [32], their cooling requirements to achieve superconductivity typically imply a large and complex measurement setup.

A different approach to localization assisted by infrastructure is presented in [52]. Hashi et al. investigate a system for locating a purely passive marker node consisting of an LC-circuit tuned to resonance at 175 kHz. The marker is driven by an external coil, and the field contributions from both the driving coil and the marker induce voltages in a regularly arranged sensor array of 25 pickup coils. In order to extract the marker's contribution, which is subsequently used for localization, the system requires a prior calibration measurement of the induced voltages with the marker absent.

6.2 Principles of Circuit Based Localization

The primary drawback of many of the existing localization schemes presented in the previous section with respect to their implementation in inductively coupled microsensor networks is the inherent requirement of complexity stemming from 3D antenna geometries, additional

---

[1]As magnetometers can also sense magnetic fields at zero frequency, the magnetic source is a permanent magnet in some cases.

infrastructure, or elaborate circuitry on the agent nodes. We therefore develop a novel localization scheme with the goal of addressing these drawbacks, which is presented in this section.

6.2.1   Network Topology

We consider a network consisting of N inductively coupled nodes. In this network, the position of a single agent node is to be estimated, while the remaining N_A nodes act as anchor nodes. In general, all anchors are assumed to have fixed and known positions and orientations, given by the vector $\boldsymbol{\theta}_n$ for the nth anchor, while the agent is characterized by the unknown parameter vector $\boldsymbol{\theta}_0$. The network may also include secondary nodes, which act neither as anchor or agent. Due to their coupling with all other nodes, the presence of secondary nodes affects the interaction of potentially all other nodes in the network and therefore also the localization process. All nodes are considered to have circular loop antennas with identical radius r and identical number of windings ν.

While we formulate the localization scheme for arbitrary network topologies, we introduce a two-dimensional, coplanar node arrangement. More specifically we consider all nodes to be placed within a square of side length b, which is centered on the origin. Furthermore all nodes have identical orientations $\mathbf{q} = \mathbf{0}$ such that the antenna lies in the mutual plane. The resulting system setup is depicted in Fig. 6.1. We use this simple setup as the basis of our subsequent analysis as it allows for an intuitive discussion of the effects governing the localization performance. The choice of this setup reduces the originally six-dimensional parameter vectors to $\boldsymbol{\theta}_n = \mathbf{p}_n = [x_n, y_n]^{\mathrm{T}}$ for the anchors and $\boldsymbol{\theta}_0 = \mathbf{p}_0 = [x_0, y_0]^{\mathrm{T}}$ for the agent.

6.2.2   Circuit Model Representation

As motivated in Section 3.2, we rely on circuit theory for a description of the individual wireless devices as well as the inductively coupled network. The circuit of the nth anchor consists of the antenna—modeled by its self-inductance L_n in series with its ohmic loss resistance R_n—and a series capacitor to tune the anchor circuit to resonance at frequency ω_{res}, i.e. the capacitance value C_n is chosen as

$$C_n = \frac{1}{L_n \omega_{\mathrm{res}}^2}. \tag{6.2}$$

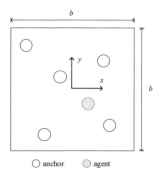

Figure 6.1. System setup for two-dimensional localization with an exemplary network arrangement.

The agent circuit is similarly modeled by the antenna elements L_0 and R_0, which are terminated in series by a generalized load impedance $Z_0(j\omega)$. The circuits of anchors and agent are depicted in Fig. 6.2.

The anchors have the capability of measuring the input impedance $Z_{\mathrm{in},n}$ at their respective open port. As there are several anchor and only a single agent node, an simple expression for $Z_{\mathrm{in},n}$ can be obtained by assuming that each anchor performs its observation based only on its pairwise interaction with the agent. This can be achieved e.g. if the anchors measure sequentially, with non-measuring anchors disconnecting their antennas from the measurement circuit. We assume that the anchors are synchronized to allow coordinated switching. This procedure eliminates any interaction between the anchors and allows formulate the observed input impedances $Z_{\mathrm{in},n}$ on the basis of a two-port model. Fig. 6.3 depicts this circuit model for the chosen implementations of anchor and agent nodes. The input impedance measured by anchor n at the discrete frequency ω_w, $w \in \{1, \ldots, N_{\mathrm{freq}}\}$, is then given by

$$Z_{\mathrm{in},nw} = R_n + j\omega_w L_n + \frac{1}{j\omega_w C_n} + \frac{\omega_w^2 M_n^2}{R_0 + j\omega_w L_0 + Z_0}, \tag{6.3}$$

where M_n is the mutual inductance between achor n and the agent. As M_n depends on the relative arrangement of anchor and agent (cf. Section 3.1.2), the mutual inductance and therefore the values $Z_{\mathrm{in},nw}$ contain information about the unknown parameters $\boldsymbol{\theta}_0$.

Fig. 6.4 shows a possible implementation of a circuit for observing the input impedance at anchor n. By using the model of an ideal current source, the current i_n, typically a sinusoid

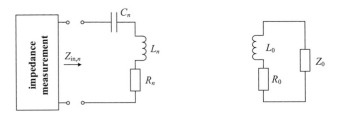

(a) Circuit model of anchor node n.　　　　(b) Circuit model of agent.

Figure 6.2. Antenna circuits of the anchor and agent nodes.

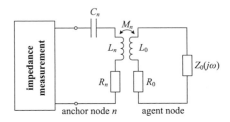

Figure 6.3. Circuit model for impedance measurements.

of frequency ω_w, can be assumed to be perfectly known. To calculate the unknown input impedance $Z_{\text{in},n}$, the voltage u_n is measured and the relation

$$Z_{\text{in},n} = \frac{u_n}{i_n} \tag{6.4}$$

is used. As with any practical measurement system, the observed voltage u_n is subject to non-systematic, random errors. As the most common source of random measurement errors for voltage observations is thermal noise, a suitable model for the noise perturbing u_n is an additive random variable which takes on values from a circularly symmetric Gaussian distribution with zero mean and variance σ_u^2. It follows that the calculated input impedance observation—denoted as O_n—is also a random variable taking on values from a circularly

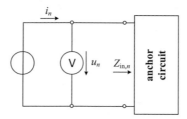

Figure 6.4. Measurement instrument model for impedance measurements.

symmetric Gaussian distribution with mean

$$\mathsf{E}\left[O_n\right] = Z_{\mathrm{in},n} \tag{6.5}$$

and variance

$$\mathsf{E}\left[|(O_n - \mathsf{E}[O_n])|^2\right] = \sigma_{\mathrm{N}}^2 = \frac{1}{i_n^2} \cdot \sigma_{\mathrm{u}}^2. \tag{6.6}$$

It should be noted that other, more practical implementations of the impedance measurement process can result in non-Gaussian distributions of the non-systematic errors in the observation. As an example, if the current i_n in the previous example would be measured by an imperfect instrument also producing an additive Gaussian error, the resulting impedance measurement would follow a ratio distribution, which is often more tedious to work with in an analytical sense.

Alternatively it would also be possible to base the localization on the measurement of other network parameters. One example is the input reflection coefficient Γ_n, which, as discussed in Section 2.2, is defined as ratio of the amplitudes of any pair of impinging wave generated by the measurement instrument and reflected wave caused by the discontinuity in line impedance presented by $Z_{\mathrm{in},n}$. The reflection coefficient of the nth anchor can be calculated as

$$\Gamma_n = \frac{Z_{\mathrm{in},n} - Z_0}{Z_{\mathrm{in},n} + Z_0}, \tag{6.7}$$

where we define Z_0, the reference impedance of the measurement instrument, to be $50\,\Omega$. Gaussianity of the measurement noise can generally not be assumed for the reflection coefficient. For the simplicity of analysis we will therefore base the localization process in the following on input impedance measurements using the above mentioned assumption of a

Gaussian distribution for the non-systematic measurement errors.

6.2.3 Optimization Process

In the following we analyze the estimation of the unknown agent position $\mathbf{p}_0$. To this end we use the dipole approximation (3.20) to model the mutual inductance between the nth anchor and the agent. We can therefore can expand the expression for the true value of the input impedance as

$$Z_{\text{in},nw} = \underbrace{R_n + j\omega_w L_n + \frac{1}{j\omega_w C_n}}_{\alpha_{nw}} + \underbrace{\frac{\omega_w^2 \mu^2 \pi^2 \nu_0^2 \nu_n^2 r_m^4 r_n^4}{16\left(R_0 + j\omega_w L_0 + Z_0\right)}}_{\beta_{nw}} \cdot \frac{J_n^2}{d_n^6}, \tag{6.8}$$

The terms summarized as α_{nw} represent the circuit of the anchor, while the values β_{nw} are related to the load of the agent on the anchor. Both quantities do not depend on the unknown parameters $\boldsymbol{\theta}_0$ and therefore contain no information that can be exploited for localization. By rotating the phase of the complex values β_{nw} to zero and subtracting the independent α_{nw}, we obtain a real valued quantity $\tilde{Z}_{nw}$. We arrange the set of all values of $\tilde{Z}_{nw}$ in a matrix $\tilde{\mathbf{Z}}(\boldsymbol{\theta}_0)$, in which the rows correspond to the anchor index n and the columns represent the frequency index w. Using the previously made assumption of all nodes having identical antennas, this matrix can be decomposed as

$$\tilde{\mathbf{Z}}(\boldsymbol{\theta}_0) = \begin{bmatrix} \chi_1(\mathbf{p}_0) \\ \chi_2(\boldsymbol{\theta}_0) \\ \vdots \\ \chi_{N_A}(\boldsymbol{\theta}_0) \end{bmatrix} \cdot \boldsymbol{\beta}^{\mathrm{T}}, \tag{6.9}$$

where we have made the definitions

$$\chi_n(\boldsymbol{\theta}_0) = \frac{J_n^2}{d_n^6} = \frac{J_n^2(\mathbf{q}_n, \mathbf{q}_0)}{\|\mathbf{p}_0 - \mathbf{p}_n\|^6}, \text{ and} \tag{6.10}$$

$$\boldsymbol{\beta}^{\mathrm{T}} = \left[|\beta_1|, \ldots, |\beta_{N_{\text{freq}}}|\right]. \tag{6.11}$$

As discussed in the previous section, the impedance measurement will inevitably introduce a random noise component which we assume to be additive i.i.d. complex Gaussian with zero mean and of variance σ_N^2, such that the observation is given by

$$\mathbf{O}(\boldsymbol{\theta}_0) = \tilde{\mathbf{Z}}(\boldsymbol{\theta}_0) + \mathbf{W}, \tag{6.12}$$

with the matrix $\mathbf{W}$ representing the samples of the measurement noise.

In the general case the maximum likelihood estimator of $\mathbf{p}_0$ and the corresponding orientation $\mathbf{q}_0$ for the observation in (6.12) is given by

$$\{\hat{\mathbf{p}}_0, \hat{\mathbf{q}}_0\} = \arg \min_{\breve{\theta}_0} \left\| \mathbf{O}(\boldsymbol{\theta}_0) - \tilde{\mathbf{Z}}(\breve{\boldsymbol{\theta}}_0) \right\|_{\mathrm{F}}^2. \tag{6.13}$$

Instead of solving this problem directly, we simplify it to gain insights into the underlying optimization problem. This can be done by expanding the cost function as

$$\left\| \mathbf{O} - \tilde{\mathbf{Z}}(\breve{\boldsymbol{\theta}}_0) \right\|_{\mathrm{F}}^2 = \sum_{n=1}^{N_A} \left\| \mathbf{O}[n,:] - \chi_n(\breve{\boldsymbol{\theta}}_0) \cdot \boldsymbol{\beta}^{\mathrm{T}} \right\|^2$$

$$= \sum_{n=1}^{N_A} \mathbf{O}[n,:] \mathbf{O}[n,:]^{\mathrm{H}} - 2\chi_n(\breve{\boldsymbol{\theta}}_0) \boldsymbol{\beta}^{\mathrm{T}} \mathbf{O}[n,:]^{\mathrm{H}} + \chi_n^2(\breve{\boldsymbol{\theta}}_0) \boldsymbol{\beta}^{\mathrm{T}} \boldsymbol{\beta}. \tag{6.14}$$

The terms $\left(\mathbf{O}[n,:] \mathbf{O}[n,:]^{\mathrm{H}} \right)$ are independent of the trial parameters $\breve{\boldsymbol{\theta}}_0$, while the terms $\left(\chi_n^2(\breve{\boldsymbol{\theta}}_0) \boldsymbol{\beta}^{\mathrm{T}} \boldsymbol{\beta} \right)$ do not depend on the observed data. We can therefore base the optimization on the sufficient statistics

$$c \cdot \mathbf{O}[n,:] \boldsymbol{\beta}, \; n \in \{1, \ldots, N_A\}, \tag{6.15}$$

where we have introduced an arbitrary real scaling factor c without loss of generality, which we choose to $c = \|\boldsymbol{\beta}\|^{-1}$. Additionally, it follows from (6.9)-(6.12) that the sufficient statistics contain real valued signal components in complex noise. We may therefore drop the imaginary part of (6.15), resulting in the observation vector

$$\bar{\mathbf{o}}(\boldsymbol{\theta}) = \boldsymbol{\chi} + \mathbf{w} = [\chi_1(\boldsymbol{\theta}_0), \ldots, \chi_{N_A}(\boldsymbol{\theta}_0)]^{\mathrm{T}} + \mathbf{w}. \tag{6.16}$$

As a result we obtain the simplified optimization problem

$$\{\hat{\mathbf{p}}_0, \hat{\mathbf{q}}_0\} = \arg \min_{\breve{\boldsymbol{\theta}}_0} \left\| \bar{\mathbf{o}}(\boldsymbol{\theta}_0) - [\chi_1(\breve{\boldsymbol{\theta}}_0), \ldots, \chi_{N_A}(\breve{\boldsymbol{\theta}}_0)]^{\mathrm{T}} \right\|^2. \tag{6.17}$$

Interestingly, for the previously introduced special case of a 2D coplanar arrangement, the terms $\chi_n(\breve{\boldsymbol{\theta}}_0)$ become identical to d_n^{-6} and the optimization in (6.17) represents a trilateration problem. It should be noted that the variance of the noise vector $\mathbf{w}$ is scaled as compared to the original noise variance σ_N^2 due to the introduced modifications of scaling the signal and neglecting the imaginary part of the noise:

$$\mathbf{w} \sim \mathcal{N}\left(\mathbf{0}, \tilde{\sigma}_N^2 \mathbf{I}\right), \qquad \tilde{\sigma}_N^2 = \frac{c^2}{2}\sigma_N^2 = \frac{\sigma_N^2}{2\|\boldsymbol{\beta}\|^2}. \tag{6.18}$$

For each element in the observation vector $\tilde{\mathbf{o}}$, we can accordingly define a measurement signal-to-noise ratio (SNR). If we define the part of $Z_{\text{in},nw}$ that depends on the agent position as the signal of interest, the SNR may be stated as

$$\text{SNR}_n = \frac{|\chi_n(\mathbf{p}_0)|^2}{\tilde{\sigma}_N^2}. \tag{6.19}$$

It follows from (6.16) and (6.18) that the SNR scales as $\|\boldsymbol{\beta}\|^2$, where the elements of $\boldsymbol{\beta}$ can be explicitly written as

$$\beta_w = \frac{\omega_w^2 \mu^2 \nu^4 \pi^2 r^8}{16\left(R_0 + j\omega_w L_0 + Z_0\right)}. \tag{6.20}$$

It is sensible to optimize $\boldsymbol{\beta}$ in order to maximize the measurement SNR. We can see from (6.20) that all values of β_w and accordingly the SNR scale with increasing frequency, the number of windings, and the antenna size. However, these parameters may be restricted in practice due to, e.g., regulations or size constraints of the antennas. We will restrict ourselves to choosing the optimum value of Z_0, which can be found as

$$Z_{0,\text{opt}} = \arg \max_{Z_0} \|\boldsymbol{\beta}\|^2. \tag{6.21}$$

For measurements with a single frequency excitation, $\|\boldsymbol{\beta}\|^2$ will be maximized in the resonant case, i.e. $Z_0 = -j\omega L_0$. This corresponds to a parallel tuning capacitor, as often used in RFID systems, and any possible additional load on the agent (chip circuitry, etc.) being open circuited. In the general case of $N_{\text{freq}} \geq 1$, the optimum choice of Z_0 will depend on the distribution of the frequency samples.

6.2.4 Performance Bound

To put the performance of the previously described estimation problem into perspective, we calculate a bound on the root mean squared error of the position estimate for networks arranged in the previously introduced 2D coplanar setup. To this end, we consider the well-known Cramér-Rao lower bound (CRLB) which states that the covariance of any unbiased estimator $\hat{\mathbf{p}}_0$ satisfies

$$\mathsf{E}\left[\left(\hat{\mathbf{p}}_0 - \mathbf{p}_0\right)\left(\hat{\mathbf{p}}_0 - \mathbf{p}_0\right)^{\mathrm{T}}\right] \succeq \mathbf{F}_{\mathbf{p}_0}^{-1}, \tag{6.22}$$

where $\mathbf{F}_{\mathbf{p}_0}$ is the Fisher information matrix of the parameter vector $\mathbf{p}_0$. Using $\mathbf{p}_0 = [x_0, y_0]^{\mathrm{T}}$, it can be calculated as [165]:

$$\mathbf{F}_{\mathbf{p}_0} = \mathsf{E}\left[\begin{array}{cc} \left(\frac{\partial \ln \Pr(\bar{\mathbf{o}}|\mathbf{p}_0)}{\partial x_0}\right)^{\mathrm{T}}\left(\frac{\partial \ln \Pr(\bar{\mathbf{o}}|\mathbf{p}_0)}{\partial x_0}\right) & \left(\frac{\partial \ln \Pr(\bar{\mathbf{o}}|\mathbf{p}_0)}{\partial x_0}\right)^{\mathrm{T}}\left(\frac{\partial \ln \Pr(\bar{\mathbf{o}}|\mathbf{p}_0)}{\partial y_0}\right) \\ \left(\frac{\partial \ln \Pr(\bar{\mathbf{o}}|\mathbf{p}_0)}{\partial y_0}\right)^{\mathrm{T}}\left(\frac{\partial \ln \Pr(\bar{\mathbf{o}}|\mathbf{p}_0)}{\partial x_0}\right) & \left(\frac{\partial \ln \Pr(\bar{\mathbf{o}}|\mathbf{p}_0)}{\partial y_0}\right)^{\mathrm{T}}\left(\frac{\partial \ln \Pr(\bar{\mathbf{o}}|\mathbf{p}_0)}{\partial y_0}\right) \end{array} \right]. \tag{6.23}$$

Here $\ln \Pr\left(\bar{\mathbf{o}}|\mathbf{p}_0\right)$ is the log-likelihood function of the observations which, given the assumed Gaussianity of the observation noise, takes the form

$$\ln \Pr\left(\bar{\mathbf{o}}|\mathbf{p}_0\right)) = \ln\left(\frac{\|\boldsymbol{\Sigma}_{\mathrm{N}}\|^{-\frac{1}{2}}}{(2\pi)^{\frac{N}{2}}}\right) - \frac{1}{2}\left\|\boldsymbol{\Sigma}_{\mathrm{N}}^{-\frac{1}{2}}\left(\bar{\mathbf{o}} - \boldsymbol{\chi}(\mathbf{p}_0)\right)\right\|^2 \tag{6.24}$$

where in this setup $\boldsymbol{\Sigma}_{\mathrm{N}} = \sigma_{\mathrm{N}}^2 \mathbf{I}/(2\|\boldsymbol{\beta}\|^2)$ is the covariance matrix of $\bar{\mathbf{o}}$ as given in (6.18). The partial derivative of the log-likelihood function with respect to x_0 is calculated as

$$\frac{\partial \ln \Pr\left(\bar{\mathbf{o}}|\mathbf{p}_0\right)}{\partial x_0} = -\frac{1}{2}\left(\frac{\partial \left(\bar{\mathbf{o}} - \boldsymbol{\chi}(\mathbf{p}_0)\right)^{\mathrm{T}}}{\partial x_0}\boldsymbol{\Sigma}_{\mathrm{N}}^{-1}\left(\bar{\mathbf{o}} - \boldsymbol{\chi}(\mathbf{p}_0)\right) + \left(\bar{\mathbf{o}} - \boldsymbol{\chi}(\mathbf{p}_0)\right)^{\mathrm{T}}\boldsymbol{\Sigma}_{\mathrm{N}}^{-1}\frac{\partial \left(\bar{\mathbf{o}} - \boldsymbol{\chi}(\mathbf{p}_0)\right)}{\partial x_0}\right)$$

$$= \left(\bar{\mathbf{o}} - \boldsymbol{\chi}(\mathbf{p}_0)\right)^{\mathrm{T}}\boldsymbol{\Sigma}_{\mathrm{N}}^{-1}\frac{\partial \boldsymbol{\chi}(\mathbf{p}_0)}{\partial x_0}, \tag{6.25}$$

and equivalently with respect to y_0. By using the fact that $\mathsf{E}\left[\left(\bar{\mathbf{o}} - \boldsymbol{\chi}(\mathbf{p}_0)\right)\left(\bar{\mathbf{o}} - \boldsymbol{\chi}(\mathbf{p}_0)\right)^{\mathrm{T}}\right] = \boldsymbol{\Sigma}_{\mathrm{N}}$, we find the Fisher information matrix as

$$\mathbf{F}_{\mathbf{p}_0} = \frac{2\|\boldsymbol{\beta}\|^2}{\sigma_{\mathrm{N}}^2}\sum_{n=1}^{N_{\mathrm{A}}}\left[\begin{array}{cc} \frac{\chi_n(\mathbf{p}_0)}{\partial x_0}\frac{\chi_n(\mathbf{p}_0)}{\partial x_0} & \frac{\chi_n(\mathbf{p}_0)}{\partial x_0}\frac{\chi_n(\mathbf{p}_0)}{\partial y_0} \\ \frac{\chi_n(\mathbf{p}_0)}{\partial y_0}\frac{\chi_n(\mathbf{p}_0)}{\partial x_0} & \frac{\chi_n(\mathbf{p}_0)}{\partial y_0}\frac{\chi_n(\mathbf{p}_0)}{\partial y_0} \end{array} \right]. \tag{6.26}$$

Due to the assumption of a 2D coplanar network arrangement, the polarization factor in (6.10) is given as $J_n = 1$ for all anchor-agent pairs. Accordingly the the derivative $\partial \chi_n / \partial x_0$ may be calculated as

$$\frac{\partial \chi_n}{\partial x_0} = -6\frac{(x_n - x_0)}{\left[(x_n - x_0)^2 + (y_n - y_0)^2\right]^4}. \tag{6.27}$$

The calculation of $\partial \chi_n / \partial y_0$ follows similarly. Inserting these derivatives into (6.25) and (6.23),

117

we obtain the Fisher information matrix

$$\mathbf{F}_{\mathbf{p}_0} = \frac{72\|\beta\|^2}{\sigma_N^2} \sum_{n=1}^{N_A} \frac{1}{((x_n - x_0)^2 + (y_n - y_0)^2)^4}$$

$$\cdot \begin{bmatrix} (x_n - x_0)^2 & (x_n - x_0)(y_n - y_0) \\ (y_n - y_0)(x_n - x_0) & (y_n - y_0)^2 \end{bmatrix}. \qquad (6.28)$$

The corresponding bound on the mean squared localization error is given by

$$\mathsf{E}\left[\|\hat{\mathbf{p}}_0 - \mathbf{p}_0\|^2\right] \geq \mathrm{tr}\left\{\mathbf{F}_{\mathbf{p}_0}^{-1}\right\}, \qquad (6.29)$$

with tr $\{\cdot\}$ denoting the matrix trace. We will use this bound in the following section both as a reference for overall localization performance and as a metric to quantify the impact of the anchor topology on the optimization.

6.3 Numerical Performance Evaluation

In this section we provide numerical analysis and evaluation of circuit based localization, again assuming a 2D coplanar arrangement. The localization process consists in solving the optimization problem generally stated in (6.13), which we achieve numerically using the Nelder-Mead simplex algorithm [105]. All involved nodes have a loop antenna with $\nu = 20$ windings, a radius of $r = 2.5$ cm, and a loss resistance of $R_0 = 1\,\Omega$. The anchor and agent nodes were matched with a series capacitance to achieve uncoupled resonance at 25 MHz. Localization is based on a single input impedance measurement per anchor, obtained at resonance frequency.

6.3.1 Evaluation of Ranging Performance

An interesting sub-problem of localization is presented by the special case of a network consisting only of a single anchor-agent pair. As we have previously assumed that all nodes are in a coplanar arrangement, it is possible to directly estimate the distance between the nodes from the impedance measurements at the anchor. The numerically evaluated ranging performance can provide insights into how useful the contributions of a single anchor are as measurement SNR decreases with range, and therefore should indicate the range limitation of the system due to the near field nature of node interaction. To this end, we define a 1D coplanar ranging setup where the position of the anchor can be assumed to be identical to the origin of the coordinate system and the position of the agent is defined by the scalar parameter $\mathbf{p}_0 = d$ representing the x coordinate, with the y coordinate set to zero.

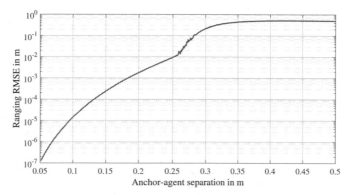

Figure 6.5. Ranging RMSE for single anchor-agent pair vs. distance.

Fig. 6.5 visualizes the achieved root mean squared error (RMSE) of the range estimates, when the true distance of the anchor was varied from 5.1 cm to 0.5 m and the noise variance was chosen as $\sigma_N^2 = 10^{-4}\,\Omega^2$. To contextualize this value, a characterization of practically achievable noise levels in a laboratory setup is provided in Chapter 8. The distance estimates were constrained to the interval $\hat{d} \in [0.05\,\mathrm{m}, 2\,\mathrm{m}]$. It can be seen that very accurate ranging is achieved at small separations due to the large spatial gradient of the input impedance over d. In this regime the ranging RMSE increases from $0.1\,\mu\mathrm{m}$ to $1\,\mathrm{cm}$ until the ranging continually breaks down at true distances between 25 cm and 35 cm: as the distance is increased, the signal of interest, i.e. the changes in the observed input impedance, becomes asymptotically smaller and approaches the magnitude of the standard variation of the noise. Once the signal component drops significantly below the noise floor, the distance estimate becomes asymptotically independent of the true distance and represents a blind guess which only depends on the noise realization, as can be seen for distances above 35 cm.

6.3.2 Impact of Initial Position

We next verify the performance of the circuit based localization method by Monte Carlo simulations over 2000 random topology instances for a network of $N_A = 5$ anchors and a single agent. All positions are drawn from a uniform distribution over the bounding box shown in Fig. 6.1 with the constraint that the antennas cannot physically intersect, i.e. $\forall m \neq$

$n : d_{mn} > 2r$. To allow for a comparison to the CRLB derived in Section 6.2.4, we again choose the RMSE as figure of merit. The noise variance is chosen as $\sigma_N^2 = 10^{-3} \, \Omega^2$, and the RMSE is calculated over 50 noise realizations per random network.

The red curves in Figure 6.6 show the empirical cumulative distribution functions (CDFs) of the RMSE achieved by optimizing (6.17). We have used two different methods for initializing the Nelder-Mead simplex algorithm. For the dashed red line, an initial trial position was randomly drawn from a uniform distribution within the bounding box. This method shows a clear performance gap compared to the square root of the CRLB, shown in black. The main reason for this is the non-convexity of the cost function, resulting in convergence to minima not corresponding to the true position. A straightforward improvement can be found

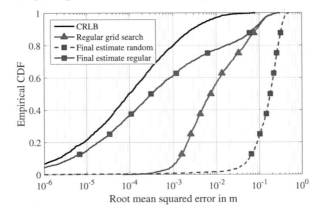

Figure 6.6. Optimization results for different initialization methods.

by performing a grid search and choosing the resulting minimum as initial point for the optimization algorithm. To this end, we evaluate the cost function on a uniform grid with step sizes of approximately 0.45 cm. The trial positions resulting from this search achieve the RMSE shown in blue, which is already better than the optimization result for random initialization. Using these trial positions to initialize the optimization algorithm leads to even further gains.

6.3.3 Reduction of Grid Search Complexity

It is clear from Fig. 6.6 that even after providing an initialization point relatively close to the true position, the performance gap to the CRLB still is significant. In an attempt to achieve even better convergence behavior, we provide a heuristic improvement of the grid search algorithm based on two insights. First, the cost function contribution of a single anchor is rotationally symmetric around the anchor's position. Furthermore, due to the d^{-6} dependency of (6.9), the total cost function is likely to be dominated by the anchor closest to the evaluated position. Both these effects can be seen in Fig. 6.7, which shows the previously used grid search for a random network topology.

The true agent position is approximately at $[-0.07, -0.13]$, while an second minimum can be seen around $[-0.17, -0.13]$. As both minima are very sharp compared to the sampling grid, the position of the sampling points will have a significant impact on the initial position and accordingly on which of the minima the subsequent optimization will converge to. This can be circumvented by increasing the grid density at expense of computational efficiency. Alternatively, based on the aforementioned insights, we can restrict the grid search to sampling the cost function in a ring around the dominant anchor. We obtain an estimate of the distance between it and the agent as

$$\hat{d}_{min} = \left(\max_n O_n \right)^{-\frac{1}{6}}, \tag{6.30}$$

with O_n being the observation of the nth anchor. We can then define a search in polar coordinates within the heuristically chosen radial limits of $d_{min} \pm 10\,\%$. The result is shown in Fig. 6.8 for the previously used topology with 12000 sample points, which is equivalent to the number of points used in the regular grid.

Using this approach, the exact position of both minima can be determined with higher accuracy due to the increased spatial resolution, and the probability of choosing an initialization point in the convex region around the true position can be increased. This can be seen from the empirical CDFs of the RMSEs shown in Figure 6.9, which demonstrate the impact of using the improved grid search to initialize optimization. Here, the dashed lines correspond to the RMSE of both the regular grid search and subsequent optimization, while the solid lines are the respective result of using the heuristically improved grid search instead. In the latter case, the performance of the optimization is fairly close to the limits imposed by the CRLB, with a median RMSE value of 126 μm. This value is conspicuously small, particularly compared to the error performance typically achieved by localization systems based on the electromagnetic far-field. However, while the result is based on strong assumptions (i.e. perfect circuit knowledge), it clearly showcases the previously mentioned benefit of the magnetic near-field having a steep spatial gradient.

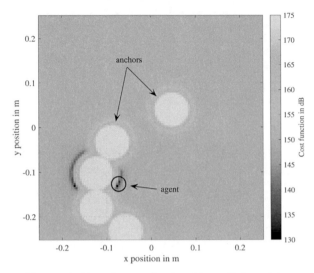

Figure 6.7. Exemplary cost function with two distinct minima.

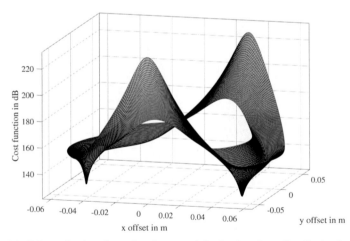

Figure 6.8. Polar evaluation of cost function around dominant anchor. Coordinates denote offset from the center the anchor.

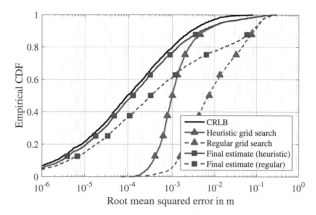

Figure 6.9. Comparison of performance for initialization with a regular (dashed) and improved (solid) grid search.

6.3.4 Optimized Choice of Anchor Topology

For planar distance-based localization techniques, conventional wisdom is that a uniform placement of the anchors on the perimeter of possible agent positions minimizes the mean CRLB of all agent positions. In [9], the authors provide a formal justification for this arrangement and derive an optimal anchor topology with slight improvement over the peripheral placement: uniformly spaced on a circle around the centroid of the bounding box. The optimum circle radius is derived as the RMS distance of all possible agent positions from the centroid. However, this result is based on an analysis assuming noisy distance observations with equal noise variance. In contrast, distance estimates reconstructed from the considered impedance measurements have low noise variance for strong coupling at close distances, with the variance growing as the anchor-agent distance increases. To account for this fact we propose using an alternative, heuristically optimized anchor topology in the shape of a cross, which aims at minimizing the average distances to a sufficient number of anchors.

In order to compare both topologies, we evaluate the CRLB for all possible positions of the agent node, i.e. small distances, for which the loop antennas of anchor and agent would intersect, were dropped. For the considered bounding box of side length b, the RMS distance to the origin of all possible positions $\mathbf{p}_0$, and therefore the optimal radius derived in [9], can be calculated by evaluating each of the statistically independent random coordinates separately,

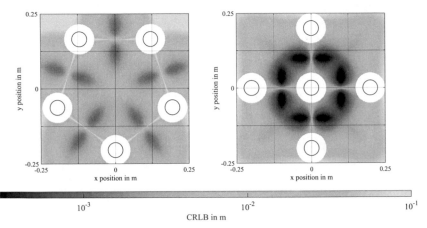

Figure 6.10. Evaluation of the Cramér-Rao lower bound on the position error variance. Left: circular anchor topology from [9]. Right: cross-shaped anchor topology.

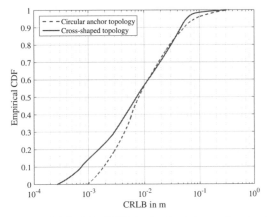

Figure 6.11. Empirical CDF of the CRLB for circular and cross-shaped anchor topology.

as follows:

$$
\sqrt{\mathsf{E}\left[\|\mathbf{p}_0\|^2\right]} = \sqrt{\mathsf{E}\left[x_0^2\right] + \mathsf{E}\left[y_0^2\right]}
$$

$$
= \sqrt{2 \cdot \int_{-b/2}^{b/2} x_0^2 \frac{1}{b} \mathrm{d}x_0}
$$

$$
= \sqrt{\frac{1}{6}b}. \tag{6.31}
$$

The results of the evaluation are shown in Fig. 6.10 as a function of the agent position and in Fig. 6.11 as empirical CDF. The median and mean value of the CRLB are slightly better for the cross-shaped topology at 7.3 mm and 1.8 cm, respectively, 8.0 mm and 2.2 cm circular topology. In addition, the cross-shaped topology indicates better behavior for unfavorable agent positions: its 99th percentile of the CRLB is 12.2 cm compared to 21.4 cm for the circular topology. As indicated in Fig. 6.10, the most unfavorable agent positions corresponding to these values are found close to the edge of the bounding box at positions dominated by the proximity of only a single anchor.

To verify the expected gains from choosing an optimized anchor topology we evaluate the achieved localization RMSE with $N_A = 5$ anchors for 2000 network topologies in which the anchors were invariably arranged in the cross-shaped topology, while the agent position was drawn from a uniform random distribution within the bounding box. The resulting errors are compared in Fig. 6.12 to those previously achieved for the same setup with a completely random network topology. The results suggest that the main benefit of the cross-shaped anchor topology lies in a higher reliability of the localization performance by preventing unfavorable anchor-agent arrangements. This becomes particularly clear when considering the 99th percentile of the RMSE as figure of merit. It is reduced from 11.5 cm by almost two orders of magnitude to a value of 2.2 mm. Based on these observations we draw the conclusion that the cross-shaped anchor setup is a well-suited choice for localization.

6.3.5 Impact of Network Density on Localization Performance

As the obtainable localization error was shown to strongly depend on the relative arrangement of anchors and agent, it can be expected that the number of used anchors N_A is a key parameter of localization performance. To judge its impact we again consider networks with completely random topology. In addition to the previously studied case with $N_A = 5$ anchors,

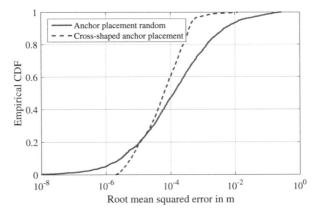

Figure 6.12. Comparison of RMS localization error for $N_A = 5$ anchors arranged randomly or in fixed in optimized positions.

we study a minimal setting with $N_A = 3$ as well as a dense network with $N_A = 10$. The empirical CDF of the RMSE achieved over 2000 random topologies is shown in Fig. 6.13. As should be expected, it can be seen that increasing the number of anchors significantly benefits the accuracy of the estimator output. This is a consequence of the increased number of available measurements as well as the greater anchor density within the bounding box. Even though using a minimal configuration of three anchors is clearly suboptimal, an RMSE of 1 cm can still be achieved with this setup for roughly 70 percent of the simulated geometries.

6.3.6 Scalability Considerations

One of the applications enabled by inductive coupling as a wireless physical layer lies in low-complexity, highly miniaturized sensor networks. It is therefore of interest to study how the proposed localization scheme is applicable to networks with nodes of various size. Similar to the scaling analysis in Section 3.4, we again consider the scaling of the entire localization system, i.e. the physical dimensions of the node antennas and the size of the bounding box are scaled simultaneously by a factor of l. An initial assessment of the impact of miniaturization on the localization performance can be obtained from the expression of the measurement SNR defined in (6.19). Using a single observation taken at the resonance frequency ω_{res} we can

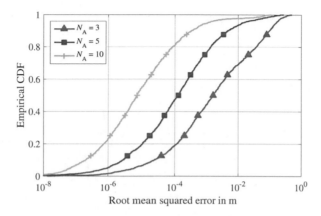

Figure 6.13. Comparison of RMSE for variations in N_A (number of anchors).

explicitly write the SNR as

$$\mathrm{SNR}_n = 2\frac{r^{16}J_n^4}{d_n^{12}R_0^2}\left(\frac{\omega^2\mu^2\nu^4\pi^2}{16}\right)^2 \cdot \frac{1}{\sigma_N^2}. \tag{6.32}$$

By definition the antenna radius and distance scale linearly with l, i.e. $r \propto l$ and $d_n \propto l$. The polarization coefficient J_n, frequency ω, and number of windings ν are independent of l. As discussed in Section 3.4, the scaling law of the agent's antenna resistance R_0 depends on the ratio between the wire radius a and skin depth δ. For simplicity we assume that the noise variance σ_N^2 is invariant with respect to a scaling of the nodes. It follows that the SNR scales as

$$\mathrm{SNR}_n \propto \begin{cases} l^4 : & \delta \ll a, \\ l^6 : & \text{otherwise.} \end{cases} \tag{6.33}$$

To evaluate the actual localization performance of a miniaturized network, we choose $l \in \{1, 1/10, 1/100\}$ which corresponds to scaling the antenna radius from 2.5 cm to 2.5 mm and finally 250 μm. The bounding box was scaled accordingly to values of $b = 50$ cm, $b = 5$ cm, and $b = 5$ mm, respectively. An insightful figure of merit to compare the localization performance in the scaled cases is the achievable positioning accuracy in relation to the overall device size. Figure 6.14 therefore shows the obtained RMSE normalized to the respective antenna radius, for a network with $N_A = 5$ anchors fixed in the previously introduced cross-shaped topology.

We have assumed that the loss resistance is fixed at $R_0 = 1\,\Omega$ over the chosen scaling range. While the low relative RMSE of the unscaled system (shown in blue) indicates that almost all

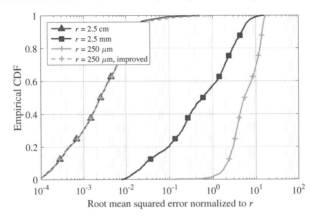

Figure 6.14. RMSE normalized to varying antenna radius r.

obtained position estimates are useful to practical applications requiring localization, the same can not be said for the miniaturized networks with $l = 1/10$ and especially $l = 1/100$, shown in red and yellow, respectively. To show that the degradation of localization performance is a consequence of the derived SNR scaling, we re-evaluate the localization performance for the smallest node size in a setting for which the SNR loss of $1/100^4$ is compensated by increasing the frequency from $f = 25\,\text{MHz}$ to $f = 100\,\text{MHz}$ and the antenna windings from $\nu = 20$ to $\nu = 100$. As expected, the relative RMSE in this configuration corresponding to the dashed yellow line, matches that of the unscaled configuration. We conclude that circuit-based near-field localization can be applied to miniaturized networks—with appropriate anchor-agent distances—if frequency and antenna windings can be chosen to guarantee sufficient SNR. However, for very small node sizes, the scaling behavior of the loss resistance additionally decreases performance.

6.4   Suboptimal Localization in 3D Space

In the previous sections, we have focused on the analysis of circuit based localization in a 2D coplanar setup, although the underlying optimization in (6.13) remains solvable in general 3D arrangements, i.e. when anchors and agent are allowed to have arbitrary positions and orientations. However, in this case the higher dimensionality of the optimization problem increase both the number of local minima of the cost function as well as required computational effort. As an alternative to jointly estimating the unknown position and orientation of the agent in a general 3D setup, it is possible to efficiently calculate an approximation of the agent position by treating the agent orientation as unknown, as shown in the following.

6.4.1 Bounding of Anchor-Agent Distance

We first discuss an upper bound on the distance of an arbitrarily placed anchor-agent pair. To this end, we solve the expression for the input impedance observed at a single anchor—as given in (6.8)—for the unknowns, i.e. the ratio of the squared polarization factor and true distance raised to the sixth power, J_n^2/d_n^6:

$$\frac{J_n^2}{d_n^6} = 16 \frac{\left(Z_{\text{in},n} - R_n - j\omega L_n - \frac{1}{j\omega C_n} \right) (R_0 + j\omega L_0 + Z_0)}{\omega^2 \mu^2 \pi^2 \nu^4 r^8} \tag{6.34}$$

As discussed in Section 3.1, the expression for polarization factor J_n as defined in (3.21) is bounded to a magnitude of $|J_n| \leq 2$. Applying this property of J_n to (6.34), we can obtain an upper bound on the true distance d_n as

$$d_n \leq d_{n,\text{max}} = \sqrt[6]{\frac{\omega^2 \mu^2 \pi^2 \nu^4 r^8}{4 \left(Z_{\text{in},n} - R_n - j\omega L_n - \frac{1}{j\omega C_n} \right) (R_0 + j\omega L_0 + Z_0)}}. \tag{6.35}$$

While the right-hand side of (6.35) contains only known values from which the maximum distance d_{max} can be calculated, we may also expand the input impedance Z_{in} using its definition in (6.8) to obtain

$$d_n \leq d_{n,\text{max}} = \sqrt[6]{\frac{4}{J_n^2}} \cdot d_n, \tag{6.36}$$

which shows how $d_{n,\text{max}}$ and the unknown true distance d_n are related by the polarization factor J_n. The term $\frac{4}{J_n^2} \geq 1$ determines the tightness of the bound.

For an intuitive interpretation of this bound one can use the fact that for an unknown anchor-agent arrangement the input impedance $Z_{\text{in},n}$ is uniquely determined by the value of k_n. This coupling coefficient can be achieved at a true distance of $d_{n,\text{max}}$ if anchor and agent are optimally coupled with a coaxial arrangement, or for distances $d < d_{n,\text{max}}$ with the misalignment of the nodes increasing as the distance decreases. It can, however, not be achieved for distances above $d_{n,\max}$ because even for optimal arrangement, the value of k_n pertaining to the observation is not achievable.

6.4.2 Impact of Noise on Distance Bounding

It needs to be pointed out that in practice the distance bounding of the agent is affected by measurement noise which perturbs the input impedance observation. As in the previous sections we assume the measured impedance $Z_{\text{in},n}$ to be subject to additive circularly symmetric complex Gaussian noise with zero mean and variance σ_N^2. Calculating the maximal distance $d_{n,\max}$ on the basis of a noisy observation $O_n = Z_{\text{in},n} + w$ accordingly results in an erroneous value $\tilde{d}_{n,\max}$, given as

$$\tilde{d}_{n,\max} = \sqrt[6]{\frac{\omega^2 \mu^2 \pi^2 \nu^4 r^8}{4\left(O_n - R_n - j\omega L_n - \frac{1}{j\omega C_n}\right)(R_0 + j\omega L_0 + Z_0)}}. \tag{6.37}$$

By again expanding the expression for the input impedance, we find the relation

$$\tilde{d}_{n,\max} = \sqrt[6]{\frac{d_{n,\max}^6}{\left(1 + \frac{d_{n}^6}{\beta J_n^2} w\right)}}, \tag{6.38}$$

where we have used the shorthand expression β defined in (6.20). One consequence of the noise w is that the expression under the root in (6.37) can become complex and therefore $\tilde{d}_{n,\max}$ can take on insensible, complex values. We further investigate this situation by equating (6.37) and (6.38) and taking the inverse:

$$\begin{aligned}
\tilde{d}_{n,\max}^{-6} &= \frac{4\left(O_n - R_n - j\omega L_n - \frac{1}{j\omega C_n}\right)(R_0 + j\omega L_0 + Z_0)}{\omega^2 \mu^2 \pi^2 \nu^4 r^8} \\
&= \frac{\left(1 + \frac{d_n^6}{\beta J_n^2} w\right)}{d_{n,\max}^6} \\
&= d_{n,\max}^{-6} + \frac{w}{4\beta}, \tag{6.39}
\end{aligned}$$

where the final step follows from using the relation in (6.36). The estimation of the true value $d_{n,\max}^{-6}$ from one or multiple noisy realizations $\tilde{d}_{n,\max}^{-6}$ can therefore be treated as the estimation of the mean of a complex Gaussian random variable. Given the knowledge that $d_{n,\max}$ is real-valued, the imaginary part of $\tilde{d}_{n,\max}^{-6}$ can be dropped and we obtain the maximum likelihood estimate of $d_{n,\mathrm{max}}^{-6}$ for N noisy observations $O_{n,i}$ as [165]

$$\hat{d}_{n,\mathrm{max}}^{-6} = \mathfrak{Re} \left\{ \frac{1}{N} \sum_{i=1}^{N} \frac{4 \left(O_{n,i} - R_n - j\omega L_n - \frac{1}{j\omega C_n} \right) (R_0 + j\omega L_0 + Z_0)}{\omega^2 \mu^2 \pi^2 \nu^4 r^8} \right\}. \tag{6.40}$$

Using these estimates to obtain $\hat{d}_{n,\mathrm{max}}$ may still lead to complex-valued distance bound estimates if the perturbation from noise leads to the real part in (6.40) becoming negative. We heuristically account for these cases by taking the absolute value of the real part:

$$\hat{d}_{n,\mathrm{max}} = \left| \hat{d}_{n,\mathrm{max}} \right|^{-\frac{1}{6}}. \tag{6.41}$$

To investigate the susceptibility of the distance bound estimation to noise as well as to gain intuition on the achieved tightness of the bound, we simulatively evaluate the distance bounding for 10000 individual anchor-agent pairs, each pair being separated by the true distance d_n and using a resonantly matched agent, i.e. $Z_0 = 1/(j\omega C_0)$. We assume random polarization factors J_n with the orientations of both nodes being randomly chosen such that all orientations are equiprobable. It has been shown in literature that for this case the probability density function of the polarization factor, $f_J(J)$, takes the form [34]

$$f_J(J) = \begin{cases} \frac{\mathrm{arcsinh}\sqrt{3}}{2\sqrt{3}} & : \quad 0 < |J| \le 1 \\ \frac{\mathrm{arcsinh}\sqrt{3} - \mathrm{arcsinh}\sqrt{J^2 - 1}}{2\sqrt{3}} & : \quad 1 < |J| \le 2 \\ 0 & : \quad 2 < |J| \end{cases}. \tag{6.42}$$

Using this distribution as well as the system parameters provided in Tab. 3.1, we calculate the input impedance values observed at the anchor of each pair. These impedances are subsequently used to obtain the distance bound for two cases: we first use noiseless impedance observations to calculate the exact distance bound $d_{n,\mathrm{max}}$ according to (6.35), and secondly obtain estimates of the true bound values from noisy input impedance observations having the noise variance $\sigma_N^2 = 10^{-3}\,\Omega^2$. The empirical CDFs of the resulting values are given in Fig. 6.15(a-c) for true distances of 10 cm, 20 cm, and 30 cm and an antenna size of $r = 2.5$ cm.

The blue solid curves in these figures are the CDFs of the noiseless case, while the red curves correspond to the noisy observations. Furthermore, the dashed solid line indicates the true distance of the anchor-agent pair on the horizontal axis. For the closest separation of 10 cm in Fig. 6.15a it can be seen that the noisy and noiseless case approximately coincide up to the 80th percentile of the CDF. This can likely be explained by the fact that, given the small separation d_n, for most orientations the SNR of the impedance observation is very high and

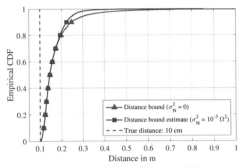

(a) Distance bound for true distance of 10 cm.

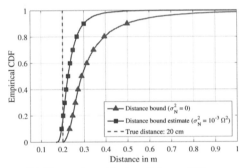

(b) Distance bound for true distance of 20 cm.

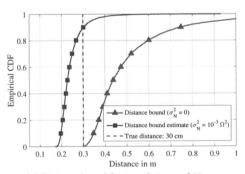

(c) Distance bound for true distance of 30 cm.

Figure 6.15. Empirical CDF of exact and estimated distance bounds for different values of the true distances.

the noise can be expected to not affect the distance bounding significantly.

When the distance is increased to 20 cm, several effects can be observed. First the empirical CDF of the true distance bound is more spread out than in the previous case, which is a consequence of the impedance observation varying with d^{-6}. Secondly the noisy observations yield estimates of the bound which are clearly biased towards shorter distances compared to the noiseless case. This effect stems from the increasingly smaller SNR of the impedance observation as the true distance increases; it can be understood by investigating the behavior of the employed estimator for the cases of very low and very high noise. Using the definition of the input impedance, the estimator in (6.41) can explicitly be formulated as

$$\hat{d}_{n,\max} = \left| \Re \left\{ \frac{1}{N} \sum_{i=1}^{N} \frac{J_n^2}{4} \cdot \frac{\left(1 + \frac{d_n^6}{\beta J_n^2} w_i\right)}{d_n} \right\} \right|^{-\frac{1}{6}}. \tag{6.43}$$

For the case of very low noise, we can assume that

$$\mathsf{E}\left[\left| \Re \left\{ \frac{d_n^6}{\beta J_n^2} \cdot w \right\} \right| \right] \ll 1, \tag{6.44}$$

for which the distance bound estimate approaches the true bound value

$$\hat{d}_{n,\max} \approx d_n \cdot \sqrt[6]{\frac{4}{J_n^2}}, \tag{6.45}$$

On the other hand, for very high noise we can assume

$$\mathsf{E}\left[\left| \Re \left\{ \frac{d_n^6}{\beta J_n^2} \cdot w \right\} \right| \right] \gg 1. \tag{6.46}$$

In this regime the distance bound takes the form

$$\hat{d}_{n,\max} \approx \left| \Re \left\{ \frac{1}{N} \sum_{i=1}^{N} \frac{w_i}{4\beta} \right\} \right|^{-\frac{1}{6}}, \tag{6.47}$$

which is independent of the true coupling between anchor and agent. For a fixed noise variance this behavior becomes more pronounced the weaker the true coupling between anchor and agent is. As can be seen in Fig. 6.15c the estimates of the bound are effectively identical for true distances of 20 cm and 30 cm.

6.4.3   Location Estimation from Distance Bounding Property

Using the distance bounding property of the input impedance measurements, we can formulate a bound on the agent position in a network with N_A anchor nodes: Each anchor's impedance measurement $Z_{\text{in},n}$ corresponds to a maximum distance such that the agent can be assumed to lie within a volume bounded by a sphere of radius $d_{n,\text{max}}$ around the nth anchor. Taking all N_A impedance measurements into account, the agent position lies within the volume formed by the intersection of the volumes spanned by all N_A balls. This region can either be used to directly estimate the agent position—e.g. by choosing the center of gravity of the resulting volume—or to limit the search space of the optimization in a 5-DOF localization problem.

The estimates $\hat{d}_{n,\text{max}}$ resulting from noisy impedance observations can be reliably employed to formulate this bound on the agent position as long as the agent is in a position such that all anchors operate in the high SNR regime specified in (6.44). As the SNR is decreased, the possibility exists that the produced estimates $\hat{d}_{n,\text{max}}$ are smaller than the true distances d_n, as can be observed in Figs 6.15(b-c). In this case the intersection volume spanned by the distance bound estimates may not contain the true agent position or be undefined, i.e. when the distance bound estimates lead to non-intersecting balls. Given that the SNR is high enough for the distance bound estimation to not degrade completely, this situation could be avoided at the expense of accuracy by increasing the estimated values $\hat{d}_{n,\text{max}}$ by a tolerance margin prior to calculating the intersection.

6.5   Conclusions

With circuit-based localization, we have introduced a novel near-field localization scheme capable of locating purely passive devices. As such, the proposed localization scheme can not only be applied to microsensor networks but also in established systems requiring localization of purely passive inductively coupled devices. A prominent example is the localization of 1-bit RFID tags.

Critically assessing the applicability of the proposed localization scheme to inductively coupled microsensor networks, we note that the requirement of having multiple anchors introduces a substantial amount of complexity into a network assumed to be comprised of low-complexity sensor nodes. This is particularly significant considering the connectivity limitations present in inductively coupled systems: if the agent is allowed to move within a large space or to have arbitrary orientations, a large number of anchors must be distributed throughout the network to guarantee a sufficient number of impedance measurements having a useful anchor-agent coupling. We will address this issue in the next chapter by showing that using passive secondary nodes with known location information does not only extend the useful range of anchor-agent interaction, but can also reduce the number of measuring anchors required to obtain an unambiguous estimate of the agent location to a single device.

In addition, we note that the localization results presented in this chapter have been based on the assumptions of perfect circuit knowledge as well as the noise being additive Gaussian, conditions that are unlikely to be fulfilled in a real-world localization system. We therefore investigate the performance of an experimental implementation of circuit-based localization in Chapter 8.

7

Single-Port Localization Systems

In the previous chapter a circuit-based near-field localization scheme has been introduced, which we have demonstrated to be suitable for application in inductively coupled networks. We have assumed that these networks consist of multiple anchor devices performing measurements of circuit parameters such as their antenna input impedance, as well as a single agent to be located on the basis of these measurements. In this chapter we will extend the circuit based localization approach to networks exhibiting only a single measurement device—and therefore only a single port of the underlying circuit model, at which measurements are obtained. The single-port setup is motivated by the typical design of inductively coupled networks: as discussed in Chapter 1, inductive coupling as a physical layer allows for an asymmetry in device complexity in the sense that the network consists of a single, highly complex central device, communicating with many low-complexity nodes. Therefore localization is also of interest for networks containing only a single anchor with measurement capabilities.

A single anchor-agent pair is, however, not sufficient for obtaining an unambiguous estimate of the agent's position. Instead, localization with a single measuring anchor is enabled by the presence of secondary nodes which interact with both the anchor and agent nodes. These secondary nodes may serve several functions within the network. Being purely passive devices, they potentially remove positional ambiguity in the agent position by acting as passive anchors. The presence of secondary nodes is also demonstrated to improve localization for weakly coupled anchor-agent pairs, similar to the magneto-inductive range extension for communication purposes discussed in Chapter 4.

Furthermore these secondary nodes may also be sensor nodes that transmit information which is not necessarily relevant to the localization. A viable scheme to perform data transmission in inductively coupled networks is the technique of load modulation discussed in Section 3.2. We propose to reuse the underlying load switching mechanism and show that it can be can be beneficially utilized to improve the achievable localization performance. Parts of this chapter have been published in [143].

7.1 System Model for Single-Port Localization

To understand the key problem of single-port localization, we briefly consider the circuit-based localization scheme discussed in the previous chapter, but with only a single anchor available to locate the unknown agent position $\mathbf{p}_0$. Using a 2D coplanar node arrangement, the measurement of the input impedance Z_{in} at the anchor is uniquely determined by the anchor-agent distance d. The localization cost function in (6.13) then has infinitely many zeros corresponding to possible agent positions. Given a noiseless measurement at the anchor, these positions lie on a circle with radius d around the anchor position. If we allow for the full degrees of freedom in the node arrangement, the solution space of the position estimate $\hat{\mathbf{p}}_0$ lies within a sphere with the radius calculated according to (6.35)[1]. While such a two-node localization system produces no unique estimate, it is possible to assign probabilities to the individual solutions $\hat{\mathbf{p}}_0$ by taking prior knowledge of the joint distribution of the true parameter vector $\boldsymbol{\theta}_0$ into account. Nevertheless, such a probabilistic approach is not expected to yield localization accuracies meeting the stringent requirements imposed by many sensor network applications. We therefore introduce the concept of passive anchors to resolve the ambiguities in the localization cost function.

While we only use a single measuring anchor with parameters $\boldsymbol{\theta}_1$, we introduce N_{S} secondary nodes with positions and orientations $\boldsymbol{\theta}_2, \ldots, \boldsymbol{\theta}_{(N_{\text{S}}+1)}$. If these parameters are known, the secondary nodes may be referred to as passive anchors. The ambiguity reduction of the passive anchors is achieved by the agent node imposing a load both directly on the anchor as well as on the passive anchors, which in turn present a load to the anchor which depends on the agent position. Because the impedance observation at the anchor depends on the joint coupling of all nodes present in the system, the notion of using pairwise anchor-agent couplings used in the previous chapter is not applicable in this setting. We therefore extend the two-port circuit model describing the inductive coupling to a multiport description for the entire network.

The measuring anchor uses the circuit shown in Fig. 7.1a. This circuit differs from the anchors considered in Chapter 6 only by the addition of a large parallel resistance R_1 which is necessary in order to describe the complete network in the Z-parameter domain. The agent and secondary nodes, on the other hand, consist of the agent circuit introduced in the previous chapter, amended by a switchable load impedance Z_{L} as shown in Fig. 7.1b. For a setup with a single anchor, single agent, and N_{S} passive anchors, the circuit describing all

---

[1] The radius of this sphere can become infinite for the case that the anchor and agent antennas are arranged such that their coupling coefficient becomes zero.

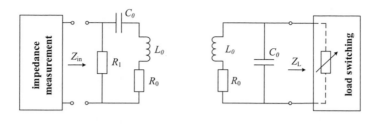

(a) Anchor node circuit.   (b) Agent / secondary node circuit.

Figure 7.1. Antenna circuits of the anchor and agent/secondary nodes.

mutual interaction is schematically depicted in Fig. 7.2. This network can be divided into the three connected multiports which we describe by their impedance matrices. The coupling impedance matrix $\mathbf{Z}_C$ (cf. (3.47)) models the inductances and loss resistances of the coupled antennas. The matching networks of all nodes are summarized in $\mathbf{Z}_M$, and $\mathbf{Z}_L$ represents the load impedances at all nodes besides the measuring anchor. The matrices $\mathbf{Z}_M$ and $\mathbf{Z}_L$ are fully known, which follows from the assumption of perfect circuit knowledge, while $\mathbf{Z}_C$ depends on the unknown agent position and orientation. Specifically, the agent parameter vector $\boldsymbol{\theta}_0 \in \mathbb{R}^6$ maps to a vector $\mathbf{k} \in [-1, 1]^{(N_s+1)}$ containing the true values of the pairwise coupling coefficients between agent and all other nodes. This mapping is may be non-injective, i.e. there may exists vectors $\tilde{\boldsymbol{\theta}}_0 \neq \boldsymbol{\theta}_0$ such that $\mathbf{k}(\tilde{\boldsymbol{\theta}}_0) = \mathbf{k}(\boldsymbol{\theta}_0)$. The coupling vector uniquely determines the matrix $\mathbf{Z}_C$ and therefore also the input impedance Z_{in}. To calculate its value, we realize that Z_{in} is the input impedance of $\mathbf{Z}_M$ at the port of the measuring anchor, when the remaining ports of $\mathbf{Z}_M$ are terminated by their respective counterparts of $\mathbf{Z}_C$ and $\mathbf{Z}_L$ as indicated in Fig. 7.2.

Without loss of generality, the input port of the measurement anchor into $\mathbf{Z}_M$ can be assigned the port index 1, followed by the ports connected to the switchable loads and finally the ports connected to the antennas. Then, by referring to the notion of a partly terminated matrix (cf. (2.16)) and by jointly treating $\mathbf{Z}_L$ and $\mathbf{Z}_C$ as load matrix of $\mathbf{Z}_M$, we obtain the true value of the input impedance as

$$Z_{\text{in}} = \mathbf{Z}_{M,11} - \mathbf{Z}_{M,12} \left(\mathbf{Z}_{M,22} + \begin{bmatrix} \mathbf{Z}_L & 0 \\ 0 & \mathbf{Z}_C \end{bmatrix} \right)^{-1} \mathbf{Z}_{M,21}. \tag{7.1}$$

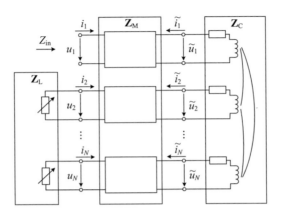

Figure 7.2. Multiport circuit model of inductively coupled network. Input impedance measurements are obtained at the anchor device.

Given the node circuits in Fig. 7.1, the elements of $\mathbf{Z}_M$ can be calculated as

$$[\mathbf{Z}_M]_{1,1} = R_1, \tag{7.2}$$

$$[\mathbf{Z}_M]_{1,(N+1)} = [\mathbf{Z}_M]_{(N+1),1} = R_1, \tag{7.3}$$

$$[\mathbf{Z}_M]_{(N+1),(N+1)} = R_1 + \frac{1}{j\omega C_0}, \tag{7.4}$$

$$[\mathbf{Z}_M]_{n,n} = [\mathbf{Z}_M]_{(n+N),(n+N)} = \frac{1}{j\omega C_0}, \tag{7.5}$$

$$[\mathbf{Z}_M]_{n,(n+N)} = [\mathbf{Z}_M]_{(n+N),n} = \frac{1}{j\omega C_0}, \tag{7.6}$$

and other entries of $\mathbf{Z}_M$ are identical to zero. We have furthermore used $N = N_S + 2$ and $n \in \{2, \ldots, N\}$.

As previously defined, the load matrix $\mathbf{Z}_L$ can change due to load switching. We assume all nodes are synchronized to allow for coordination of the switching process, and that all nodes switch their loads simultaneously. If we therefore divide time into discrete states $s \in$

$\{1, \ldots, S\}$, $\mathbf{Z}_L$ is characterized during state s by the current values of all load impedances as

$$\mathbf{Z}_L^{(s)} = \begin{bmatrix} Z_{L,1}^{(s)} & & 0 \\ & \ddots & \\ 0 & & Z_{L,(N_S+1)}^{(s)} \end{bmatrix}. \tag{7.7}$$

Herein we arbitrarily define the load impedance index $(N_S + 1)$ to correspond to the load of the agent node. The load matrices $\mathbf{Z}_L^{(s)}$ can either be unknown, in the sense that they represent messages transmitted by the respective nodes, or assumed to be known a priori for each state, similar to the concept of a training sequence.

The input impedance $Z_{in}(\mathbf{k}(\boldsymbol{\theta}_0))$ can therefore be measured for S different states of the load matrix and at N_{freq} different frequencies ω_w. Each of these observations is subject to noise, which we model as additive i.i.d. complex symmetric Gaussian with zero mean and variance σ_N^2:

$$O_w^{(s)} = Z_{in,w}^{(s)} + N_w^{(s)}, \quad N_w^{(s)} \sim \mathcal{CN}(0, \sigma_N^2). \tag{7.8}$$

An estimate of the agent position and orientation can then be found at the minimum of the total squared impedance mismatch as

$$\{\hat{\mathbf{p}}_0, \mathbf{q}_0\} = \arg \min_{\check{\boldsymbol{\theta}}_0} \sum_{s=1}^{S} \sum_{w=1}^{N_{freq}} \left| O_w^{(s)} - Z_{in,w}^{(s)}(\check{\mathbf{k}}(\check{\boldsymbol{\theta}}_0)) \right|^2. \tag{7.9}$$

7.2 Single-Port Localization Using Passive Anchors

One can expect two effects from the presence of passive anchors in circuit-based near-field localization: the reduction of localization ambiguities due to additional location information and an extension of the usable range of the measuring anchor due to the joint coupling of an anchor-agent pair with all passive anchors. In the following we provide a justification for both effects before studying the actual performance of single-port localization. We once again restrict ourselves to the 2D coplanar case for the sake of simplicity.

If not specified further, we assume all antennas of the wireless devices in this section to have identical specifications with an antenna loop radius of 2.5 cm, an inductance of $L_0 = 3.7 \, \mu H$, and a purely ohmic loss resistance of $R_0 = 4.5 \, \Omega$. The matching networks of all nodes are designed to achieve resonance at $f_{res} = 25.4 \, MHz$ with $C_0 = 10.6 \, pF$ and $R_1 = 1 \, M\Omega$. Unless

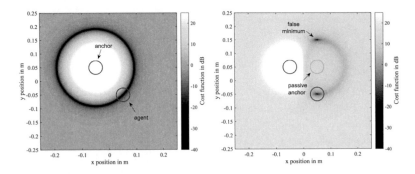

(a) Single anchor-agent pair. **(b)** Anchor-agent pair with passive anchor.

Figure 7.3. Localization cost function in dB for $\sigma_N^2 = 0$. The measuring anchor antenna is indicated in red, the agent antenna in blue, and passive anchor positions in green. Antennas are drawn to scale.

noted otherwise the switchable loads have a default value of $Z_L = 10\,\mathrm{M\Omega}$ to numerically emulate an open circuit.

7.2.1 Reduction of Localization Ambiguities

Using the system model introduced in the previous section, we can provide intuition on how both the presence of passive anchors and load switching can impact localization performance by affecting the nature of the underlying optimization problem in (7.9). To this end, we begin by again considering a single anchor-agent pair arranged in a 2D coplanar configuration with agent position $\mathbf{p}_0 = [0.05, -0.05]^T$ and anchor position $\mathbf{p}_1 = [-0.05, 0.05]^T$. The resulting noiseless cost function over all possible trial positions $\check{\mathbf{p}}_0$ is visualized in Fig. 7.3a. All antenna radii were chosen to $r = 2.5\,\mathrm{cm}$ and are drawn to scale with the anchor antenna shown in red and the agent antenna in blue. As expected, the circular symmetry of the coupling leads to infinitely many trial positions $\check{\mathbf{p}}_0$ which produce the true value of k, hence leading to a ring-shaped minimum of the cost function.

In Fig. 7.3b, a single passive anchor has been added to the network at the position $\mathbf{p}_2 = [0.05, 0.05]^T$. This configuration results in two distinct minima, one being at the true agent position, and one being mirrored along the axis connecting the measuring and passive anchor. In fact, this type of ambiguity occurs for any 2D coplanar configuration of passive anchors which exhibits symmetry with respect to the measuring anchor.

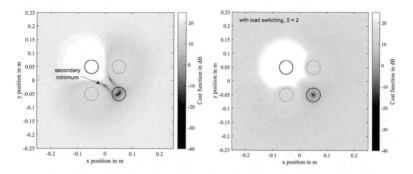

(a) Anchor-agent pair with two passive anchors.

(b) Anchor-agent pair with two passive anchors and additional load switching.

Figure 7.4. Localization cost function in dB for $\sigma_N^2 = 0$. The anchor antenna is indicated in red, the agent antenna in blue, and passive anchor positions in green. Antennas are drawn to scale.

Symmetry implies the existence of positions $\mathbf{p}_0$, $\tilde{\mathbf{p}}_0$ which fulfill the condition

$$\mathbf{k}(\mathbf{p}_0) = \mathbf{k}(\tilde{\mathbf{p}}_0), \qquad \mathbf{p}_0 \neq \tilde{\mathbf{p}}_0, \tag{7.10}$$

leading to multiple cost function minima. The ambiguity resulting from (7.10) is eliminated through the addition of another passive anchor to the network at position $\mathbf{p}_3 = [-0.05, -0.05]^T$, as depicted in Fig. 7.4a. However, there are secondary minima visible close to the true agent position. These minima can occur because the anchor's input impedance measurement, which maps the two-dimensional position space of the agent to a point in the complex impedance plane, is not guaranteed to be unique for each agent position. More specifically, multiple minima occur if there exist positions $\mathbf{p}_0$, $\tilde{\mathbf{p}}_0$ with distinct coupling vectors $\mathbf{k} \neq \tilde{\mathbf{k}}$ which lead to the same impedance observation, i.e. $Z_{\text{in}}(\mathbf{k}(\mathbf{p}_0)) = Z_{\text{in}}(\tilde{\mathbf{k}}(\tilde{\mathbf{p}}_0))$. This condition can formally be stated as

$$\left(\mathbf{Z}_{\text{M},22} + \begin{bmatrix} \mathbf{Z}_L & \mathbf{0} \\ \mathbf{0} & \mathbf{Z}_C(\tilde{\mathbf{k}}) \end{bmatrix} \right)^{-1}_{xy} [\mathbf{Z}_{\text{M},12}]_x [\mathbf{Z}_{\text{M},21}]_y =$$
$$- (\mathbf{Z}_{\text{in}}(\mathbf{p}_0) - \mathbf{Z}_{\text{M},11}) - \sum_i \sum_j (1 - \delta_{ix}\delta_{jy})$$

143

$$\left[\mathbf{Z}_{\mathrm{M},12} \left(\mathbf{Z}_{\mathrm{M},22} + \begin{bmatrix} \mathbf{Z}_{\mathrm{L}} & 0 \\ 0 & \mathbf{Z}_{\mathrm{M}}(\tilde{\mathbf{k}}) \end{bmatrix} \right)^{-1} \mathbf{Z}_{\mathrm{M},21} \right]_{ij} , \qquad (7.11)$$

where δ_{mn} is the Kronecker delta function

$$\delta_{mn} = \begin{cases} 0 & : \quad m \neq n, \\ 1 & : \quad m = n. \end{cases} \qquad (7.12)$$

It should be noted that the previous examples evaluate the cost function on the basis of a single impedance measurement, i.e. $S = 1$ and $N_{\mathrm{freq}} = 1$, where the load impedance was set to $Z_{\mathrm{L}} = 10\,\mathrm{M\Omega}$ at all passive anchors and the agent. In a next step, the cost function ambiguity of the configuration with two passive anchors is fully eliminated by using two measurements while employing load switching between observations. Specifically, we choose the loads in each state such that only one of the passive anchors experiences a strong interaction with the agent and measuring anchor. To this end we note that for the default load impedance of $10\,\mathrm{M\Omega}$, the passive anchors achieve uncoupled resonance at the design frequency. However, when setting a passive anchor load to $0\,\Omega$, the respective circuit is detuned as the capacitance is canceled, resulting in the passive anchor interacting more weakly with the remaining nodes in the network. We follow the intuition that by detuning both passive anchors alternately and superimposing the resulting cost functions of both states, the positions of the false cost function minima occur at different positions for each state, while the desired minimum is always at the true agent position and superimposes constructively. We verify this notion by selecting the load matrices $\mathbf{Z}_{\mathrm{L}}^{(1)} = \mathrm{diag}\,\{0\,\Omega, 10\,\mathrm{M\Omega}, 10\,\mathrm{M\Omega}\}$ and $\mathbf{Z}_{\mathrm{L}}^{(2)} = \mathrm{diag}\,\{10\,\mathrm{M\Omega}, 0\,\Omega, 10\,\mathrm{M\Omega}\}$. As expected the resulting cost function, shown in Fig. 7.4b, exhibits a unique and clearly defined minimum. It should be noted that switching the load of the agent node would lead to observations for which the agent is detuned and only weakly influences the measurement. We therefore always select a value of $10\,\mathrm{M\Omega}$ for the agent load impedance.

7.2.2 Range Estimation Assisted by Passive Anchors

To investigate the impact of passive anchors on the usable measurement range, we revisit the 1D ranging problem studied in Section 6.3. Building on the concept of passive relays providing range extension for communicating nodes as discussed in Chapter 4, we initially add a single passive anchor to the setup to act as relay and evaluate the ranging performance. Its load impedance Z_{R} is chosen to achieve resonance at f_{res} for an uncoupled state.

The achieved root mean squared error (RMSE) for ranging supported by a single passive anchor is shown in Fig. 7.5. Herein the black dashed curve corresponds to the previously investigated case without a passive anchor. Noting that this setup yields satisfying results until a true anchor-agent distance of approximately 26 cm with an RMSE below 1 cm, we choose to position the passive anchor close to this breakdown point. Specifically we first evaluate the position $[20\,\mathrm{cm}, 6\,\mathrm{cm}]^\mathrm{T}$ for the secondary node. The corresponding RMSE is visualized by the

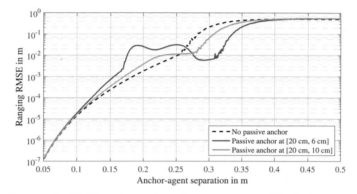

Figure 7.5. Ranging RMSE vs. distance for anchor-agent pair assisted by single passive anchor.

magenta curve in Fig. 7.5. Compared to the case without secondary nodes, this anchor position increases the distance at which ranging begins to break down from approximately 26 cm to 32 cm. On the other hand, the ranging performance at distances close to the passive anchor is significantly degraded. We conjecture that the agent experiences detuning in proximity of the passive anchor and investigate a second placement for the secondary node at $[20\,\mathrm{cm}, 10\,\mathrm{cm}]^\mathrm{T}$, which leads to the RMSE shown by the green curve. In the latter case, the performance degradation close to the passive anchor position is clearly reduced, resulting in an RMSE on the order of 1 cm for distances between the measuring anchor and the agent of up to 28 cm. An alternative strategy to improve the ranging performance using a passive anchor could be found in optimizing the load impedance Z_R.

Motivated by the observed ranging improvement from a single passive anchor, we next investigate if the maximum usable distance of the ranging process can be extended even further using a large number of passive anchors. Specifically we arrange $N_\mathrm{S} = 35$ passive anchors along the y axis with the intuition of achieving a stronger interaction of anchor and agent due to the magnetoinductive waveguide effect described in Chapter 4. Two possible

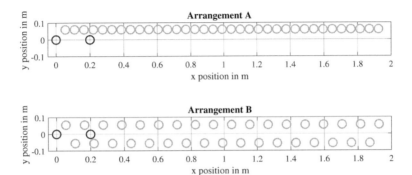

Figure 7.6. Possible arrangements of passive anchors (green) to extend ranging performance of anchor-agent pair (red and blue, respectively). The antennas are drawn to scale.

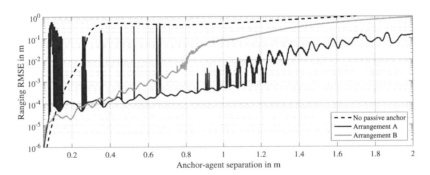

Figure 7.7. Ranging RMSE vs. distance for anchor-agent pair assisted by multiple passive anchors.

arrangements are shown in Fig. 7.6 for an exemplary anchor-agent distance of 20 cm. Here the red, blue, and green circles represent the loop antennas of the measuring anchor, agent, and passive anchors, respectively.

The top configuration, which we denote as *arrangement A*, places the passive anchors on a single line at $y = 6$ cm with a separation of 5.5 cm between the individual antennas to maximize their mutual coupling. The difference of *arrangement B*, which is shown on the bottom, is a sign flip of the y coordinate of every other passive anchor.

The impact of the arrangements on the RMSE ranging performance can be seen from the simulation results in Fig. 7.7. Both arrangements achieve good ranging RMSE at distances where the estimation breaks down for the investigated configurations using a single passive anchor. The RMSE behavior of arrangement A is particularly interesting. It exhibits a sub-millimeter baseline error for distances up to approximately 1.2 m, but shows several peaks at distinct positions for which the RMSE is drastically increased, even at very small true distances. The reason for this behavior can be found by closer examination of the cost function for the positions showing the performance degradation.

Fig. 7.8 depicts the cost function of arrangement A as a function of the trial distance $\tilde{d}$, given a noisy realization of the anchor measurement. The true distance of the agent is $d = 9.8$ cm for the blue curve and $d = 20$ cm for the red curve. The cost function is clearly

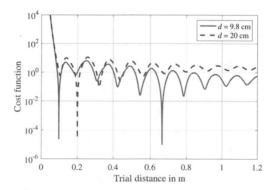

Figure 7.8. Cost function vs. trial distance using arrangement A for different true distances.

non-convex with several local minima. These local minima are most likely a result of the quasiperiodic arrangement of the passive anchors, as the separation between the cost function minima matches twice the distance between two passive anchor nodes. For $d = 9.8$ cm, which is one of the degraded positions, the cost function exhibits two dominant minima, one at the true distance and one at $\tilde{d} = 66.53\,\text{cm}$. On the other hand, the cost function for a true distance of 20 cm only has one dominant minimum leading to the observed small RMSE at this position.

It can be shown by numerical analysis that the mapping of $d \to Z_{\text{in}}$ is unique for arrangement B but not for arrangement A, i.e. there exist $\tilde{d} \neq d$ s.t. $Z_{\text{in}}(\tilde{d}) = Z_{\text{in}}(d)$ in arrangement A. Using arrangement B therefore yields a gradual increase of the ranging RMSE with the error performance being sub-millimeter up to 60.4 cm and sub-centimeter up to 80.1 cm of range.

7.2.3 Simulative Evaluation of Localization Performance

To evaluate the performance of the single-port localization scheme, the localization error was studied by numerically solving the optimization in (7.9). We consider 500 random network topologies where a measuring anchor, a single agent, and multiple passive anchors were arranged in a 2D coplanar fashion with random positions chosen uniformly within a bounding box with $b = 0.5\,\text{m}$, such that the position estimate can be bounded as $\|\mathbf{p}_0\|_\infty \leq b/2$. All RMSE values were evaluated over 50 noise realizations.

We first investigate the impact of load switching on localization performance. We therefore evaluate the absolute localization error obtained for different numbers of load switching states S, using a network with $N_S = 10$ passive anchors. To this end the load impedances are pseudo-randomly chosen as $Z_{\text{L},n}^{(s)} = 0\,\Omega$ or $Z_{\text{L},n}^{(s)} = 10\,\text{M}\Omega$ with equal probability independently for each anchor n and time slot s. For the time being we assume that the resulting load sequences $\mathbf{Z}_{\text{L},n}^{(1)}, \ldots, \mathbf{Z}_{\text{L},n}^{(S)}$ are training sequences that are known a priori and can therefore be assumed as given when solving the optimization problem (7.9).

Fig. 7.9 shows the empirical cumulative distribution function (CDF) of the localization RMSE for different numbers of load states S. The black dashed line indicates the performance for blind guessing and is therefore an upper bound on the error performance, while the blue and red dashed lines show the CRLB of the localization error corresponding to the cases of $S = 1$ and $S = 20$, respectively. For this simulation, the input impedance was only measured at the resonance frequency ω_{res}, i.e. $N_{\text{freq}} = 1$. It can be observed that for the case of no load switching ($S = 1$), a median RMSE of about 14.9 cm is achieved. This result suggests that localization with limited accuracy is possible even if the anchor obtains only a single impedance measurement. However, load switching drastically improves the error performance: with $S = 20$ a median RMSE of 8.4 mm is achieved. It is important to note that

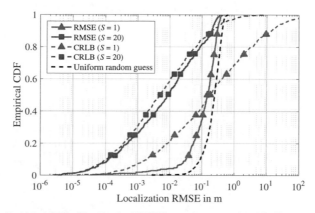

Figure 7.9. Empirical CDF of localization RMSE for random networks with $N_S = 10$ and $N_{\text{freq}} = 1$. The noise variance was set to $\sigma_N^2 = S \cdot 10^{-4} \, \Omega^2$.

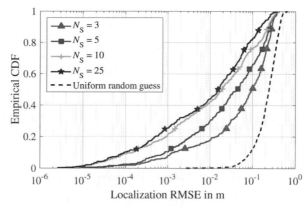

Figure 7.10. Empirical CDF of localization RMSE with variable number of passive anchors with $N_{\text{freq}} = 1$, $S = 5$, and $\sigma_N^2 = S \cdot 10^{-4} \, \Omega^2$.

the noise variance was chosen to scale with the number of measurements as $\sigma_N^2 = S \cdot 10^{-4} \, \Omega^2$. This means that the observed improvement is not a result of noise averaging over multiple measurements, but solely due to the reduction of ambiguities in the underlying cost function. The behavior of the CRLB should also be addressed, as the RMSE is partly below the CRLB, most notably for the case with $S = 1$. This is a result of the bias of the location estimate imposed by constraining the agent position to a bounding box, while the CRLB is only a lower bound on the error for an unbiased location estimate.

The achievable localization performance also depends on the number of passive anchors used in the network, as indicated in the results shown in Fig. 7.10. Herein, the number of passive anchors was varied while using load switching with $S = 5$ states and $N_{\text{freq}} = 1$. It can be seen that the error generally decreases with an increasing number of passive anchors. While the scenarios with $N_S = 3$ passive anchors achieve a median RMSE of approximately 9.8 cm, using $N_S = 25$ anchors yields a median RMSE of 1.5 cm. This improvement can be attributed to the near-field nature of the interaction between the nodes. For a distinct cost function minimum, the agent requires significant coupling to multiple passive anchors nodes, which in turn need to interact with the anchor. As demonstrated previously, dense configurations of passive anchors can lead to an increased space of true agent positions at which the observation of the measuring anchor leads to a reliable position estimate.

As next step we investigate localization with imperfect circuit knowledge. The previously made assumption of perfect circuit knowledge, i.e. knowledge of the matrices $\mathbf{Z}_M$ and $\mathbf{Z}_L$ is difficult to obtain in practical systems due to e.g. manufacturing tolerances or aging effects. For a typical hardware realization of the node circuitry, tolerances may exceed 10 % of the nominal value of the circuit element. To cope with this limitation, one possibility is to utilize a set of a priori calibration measurements for the localization, as discussed in Chapter 8. Alternatively, if the underlying circuit structure is known but the precise values of the circuit elements are not, it is feasible to treat the unknown circuit elements as nuisance parameters. The optimization problem then becomes

$$\left\{ \hat{\mathbf{p}}_0, \check{\boldsymbol{\lambda}} \right\} = \arg\min_{\check{\mathbf{p}}_0, \check{\boldsymbol{\lambda}}} \sum_{s=1}^{S} \sum_{w=1}^{N_{\text{freq}}} \left| O_w^{(s)} - Z_{\text{in},w}^{(s)}(\check{\mathbf{k}}(\check{\mathbf{p}}_0), \check{\boldsymbol{\lambda}}) \right|^2 \qquad (7.13)$$

$$\text{s.t. } \|\check{\mathbf{p}}_0\|_\infty \leq b/2,$$

where $\boldsymbol{\lambda}$ denotes the vector of nuisance parameters to be estimated.

To practically evaluate (7.13) we again consider a setting with $N_S = 25$ passive anchors,

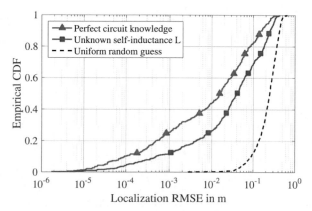

Figure 7.11. Empirical CDF of localization RMSE for unknown L_0, with $N_S = 25$, $S = 5$ and $N_{\text{freq}} = 1$.

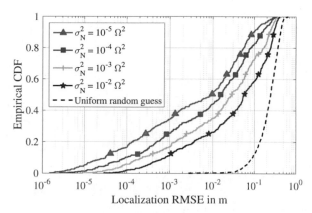

Figure 7.12. Empirical CDF of localization RMSE for different values of the noise variance σ_N^2, with $N_S = 25$, $S = 5$ and $N_{\text{freq}} = 1$.

however, the true value of the self-inductances L_0 is now unknown[2] and included as nuisance parameter in the estimation rule. The CDF of the resulting localization error for a noise variance of $\sigma_N^2 = 10^{-4}\,\Omega^2$ is compared in Fig. 7.11 to the same setup with perfect circuit knowledge. As can be expected the localization performance is decreased under the condition of imperfect circuit knowledge. However, the proposed localization scheme still produces satisfactory results with a median localization RMSE of 4.1 cm.

Finally we investigate the localization performance as a function of the noise level present at the measuring anchor. Fig. 7.12 shows the localization performance for different values of the noise variance σ_N^2 between $10^{-5}\,\Omega^2$ and $10^{-2}\,\Omega^2$. For context on these values the reader is referred to the assessment of measurement noise in an experimental setup provided in Section 8.2. For the considered network containing $N_S = 25$ passive anchors and using load switching with $S = 5$ and $N_{\text{freq}} = 1$, the median localization error clearly depends on the noise variance. However, even for the highest chosen variance setting of $\sigma_N^2 = 10^{-2}\,\Omega^2$ the achieved error performance is acceptable with a median value of 5.6 cm.

7.3 Joint Decoding and Localization

In the previous section we have considered the load states of the passive anchors to be known a priori, which would typically be the case if the wireless network is operating primarily to perform localization. However we envision localization in inductively coupled networks as a secondary application that reuses the communication infrastructure. In a more practical setup we can therefore interpret the passive anchors as nodes—with known locations—that communicate a message to a central unit, i.e. the measuring anchor, using load modulation (cf. Section 3.2). Assuming a binary load modulation scheme, the load states then represent the current bit state of each respective passive anchor, which in this context may be referred to as transmitter. The key difference of this setup lies in the fact that the load states of the transmitters are unknown and need to be estimated.

To analyze this optimization problem we again consider the localization of a single agent in a 2D coplanar setting, with the agent having a known load $Z_L = 10\,\text{M}\Omega$ and the loads of the N_S transmitters with known locations are chosen according to their transmitted bits. We refer to the bits of all transmitters at state s as a message, which is represented by a message

---

[2]It was still assumed for this simulation that the self-inductance L_0, although unknown, is identical for all nodes.

vector $\mathbf{c}^{(s)} \in \{0,1\}^{N_S}$. The message vector uniquely determines the load matrix as

$$\mathbf{Z}_L^{(s)} = \text{diag}\left\{ 10\,\text{M}\Omega \cdot \mathbf{c}^{(s)} \right\}. \tag{7.14}$$

One message is transmitted per state such that the total bits transmitted over S states are summarized in a message matrix $\mathbf{C} \in \{0,1\}^{(N_S \times S)}$. Our interested lies in estimating both the unknown agent position $\mathbf{p}_0$ as well as the messages $\mathbf{C}$. The optimal solutions for both estimates are mutually dependent and therefore estimation needs to be performed jointly. From the noisy impedance observations in (7.8) the joint optimization of the agent position $\mathbf{p}_0$ and message $\mathbf{C}$ can be formulated as the least squares minimization given by

$$\left\{ \hat{\mathbf{C}}, \hat{\mathbf{p}}_0 \right\} = \arg\min_{\check{\mathbf{p}}_0, \check{\mathbf{C}}} \sum_{s=1}^{S} \sum_{w=1}^{N_{\text{freq}}} \left| O_w^{(s)} - Z_{\text{in},w}^{(s)}\left(\check{\mathbf{k}}(\check{\mathbf{p}}_0), \check{\mathbf{c}}^{(s)} \right) \right|^2 \tag{7.15}$$

$$\text{s.t. } \|\check{\mathbf{p}}_0\|_\infty \leq b/2.$$

It can be seen that this problem is equivalent to the optimization given in (7.13) with the unknown entries of the message matrix $\mathbf{C}$ representing the nuisance parameters. As the space of possible message hypotheses is both discrete and finite, optimal joint estimation can be trivially achieved by performing localization of the agent for every message hypothesis $\check{\mathbf{C}}_j$ and choosing the pair $(\check{\mathbf{C}}_j, \hat{\mathbf{p}}_{0,j})$ with the lowest residual error of the localization as overall estimates.

The feasibility of this approach is demonstrated in Fig. 7.13, which shows the cost function and localization error obtained from the *noiseless* evaluation of the cost function in (7.15) over all possible message hypotheses for an exemplary network with $N_S = 3$ transmit nodes and a transmission length of $S = 2$. It can be seen that the message hypothesis with index $i = 58$ corresponds to a clear global minimum both for the residual cost function error as well as the localization error.

To understand the computational complexity of the optimization in (7.15), we assume the localization is performed as an exhaustive grid search with a fixed number of trial positions $\check{\mathbf{p}}_{0,i}$ where $i \in \{1, \ldots, N_P\}$. The time required to calculate the squared impedance mismatch for any single set of trial values $\check{\mathbf{p}}_{0,i}$ and $\check{\mathbf{c}}_j^{(s)}$ is denoted by τ_c, such that the summation over all states and frequencies results in the runtime of a single cost function evaluation to be $S \cdot N_{\text{freq}} \cdot \tau_c$. The exhaustive minimization over all possible combinations of $\check{\mathbf{C}}_j$ and $\check{\mathbf{p}}_{0,i}$ leads to $N_P \cdot 2^{N_S \cdot S}$ cost function evaluations, i.e. the total time for the optimization τ_{opt} is given

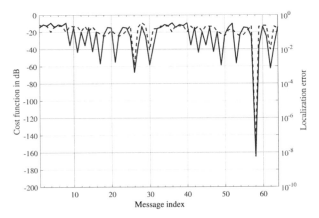

Figure 7.13. Cost function (solid) and localization error (dashed) for all message hypotheses in exemplary arrangement ($S = 2$, $N_\mathrm{S} = 3$).

by

$$\tau_\mathrm{opt} = N_\mathrm{P} \cdot 2^{N_\mathrm{S} \cdot S} \cdot S \cdot N_\mathrm{freq} \cdot \tau_\mathrm{c}. \tag{7.16}$$

It is possible to provide a more efficient reformulation of the joint decoding and localization in (7.15) by using the fact that minimization over the messages comprising the columns of $\check{\mathbf{C}}_j$ can be performed individually. The resulting optimization can be stated as

$$\left\{ \hat{\mathbf{c}}^{(1)}, \dots, \hat{\mathbf{c}}^{(S)}, \hat{\mathbf{p}}_0 \right\} = \arg\min_{\check{\mathbf{p}}_0} \sum_{s=1}^{S} \left(\min_{\check{\mathbf{c}}^{(s)}} \sum_{w=1}^{N_\mathrm{freq}} \left| O_w^{(s)} - Z_{\mathrm{in},w}^{(s)} \left(\check{\mathbf{k}}(\check{\mathbf{p}}_0), \check{\mathbf{c}}^{(s)} \right) \right|^2 \right) \tag{7.17}$$

$$\text{s.t. } \| \check{\mathbf{p}}_0 \|_\infty \leq b/2.$$

The inner minimization over the message bits of $\check{\mathbf{c}}^{(s)}$ involves $2^{N_\mathrm{S}} \cdot N_\mathrm{freq}$ evaluations of the impedance mismatch term. Further minimizing the result of the inner optimization over the N_P trial positions and summing over all states, the total runtime $\tilde{\tau}_\mathrm{opt}$ of the modified joint decoding and localization problem results in

$$\tilde{\tau}_\mathrm{opt} = N_\mathrm{P} \cdot 2^{N_\mathrm{S}} \cdot S \cdot N_\mathrm{freq} \cdot \tau_\mathrm{c}. \tag{7.18}$$

Comparing the runtimes of both variants, we find that the optimization in (7.17) is exponentially less complex with $\tau_{\mathrm{opt}} = 2^{(S-1)N_{\mathrm{S}}} \cdot \tilde{\tau}_{\mathrm{opt}}$, though it should be noted that when using non-exhaustive minimization methods, e.g. numerical gradient search, this ratio may change. The computational complexity of the optimization in (7.17) can additionally be reduced by noting that the input impedances only need to be calculated once for different states s and $\bar{s}$, i.e.

$$Z_{\mathrm{in}} \left(\check{\mathbf{k}}(\check{\mathbf{p}}_{0,i}), \check{\mathbf{c}}_j^{(s)} \right) = Z_{\mathrm{in}} \left(\check{\mathbf{k}}(\check{\mathbf{p}}_{0,i}), \check{\mathbf{c}}_j^{(\bar{s})} \right) \quad \forall \check{\mathbf{c}}_j^{(s)} = \check{\mathbf{c}}_j^{(\bar{s})}. \tag{7.19}$$

If the time to store and retrieve these values is negligible compared to the time to calculate them, the total runtime of the optimization can be decreased by a factor of S.

To justify the premise of the modified optimization in (7.17), namely that the cost function can be minimized over each of the messages $\mathbf{c}^{(s)}$ independently, we first consider the case of a single message being sent, i.e. $S = 1$. For the sake of notational clarity we assume $N_{\mathrm{freq}} = 1$, but the argument holds for arbitrary numbers of frequency samples. In this setting we can find $\Pr(\check{\mathbf{c}}_j \,|\, \check{\mathbf{p}}_{0,i}, O)$, the probability that a specific message $\check{\mathbf{c}}_j$ has been transmitted conditional on the trial position $\check{\mathbf{p}}_{0,i}$ and observation O, by using Bayes' theorem:

$$\begin{aligned} \Pr(O \,|\, \check{\mathbf{c}}_j, \check{\mathbf{p}}_{0,i}) &= \frac{\Pr(O) \cdot \Pr(\check{\mathbf{c}}_j, \check{\mathbf{p}}_{0,i} \,|\, O)}{\Pr(\check{\mathbf{c}}_j, \check{\mathbf{p}}_{0,i})} \\ &= \frac{\Pr(O) \cdot \Pr(\check{\mathbf{c}}_j \,|\, \check{\mathbf{p}}_{0,i}, O) \cdot \Pr(\check{\mathbf{p}}_{0,i} \,|\, O)}{\Pr(\check{\mathbf{c}}_j, \check{\mathbf{p}}_{0,i})}, \end{aligned} \tag{7.20}$$

where we have applied the identity $\Pr(A, B \,|\, C) = \Pr(A \,|\, B, C) \cdot \Pr(B \,|\, C)$ [113]. Solving for $\Pr(\check{\mathbf{c}}_j \,|\, \check{\mathbf{p}}_{0,i}, O)$, we find

$$\begin{aligned} \Pr(\check{\mathbf{c}}_j \,|\, \check{\mathbf{p}}_{0,i}, O) &= \frac{\Pr(O \,|\, \check{\mathbf{c}}_j, \check{\mathbf{p}}_{0,i}) \cdot \Pr(\check{\mathbf{c}}_j) \cdot \Pr(\check{\mathbf{p}}_{0,i})}{\Pr(O) \cdot \Pr(\check{\mathbf{p}}_{0,i} \,|\, O)} \\ &= \frac{\Pr(O \,|\, \check{\mathbf{c}}_j, \check{\mathbf{p}}_{0,i}) \cdot \Pr(\check{\mathbf{c}}_j)}{\Pr(O \,|\, \check{\mathbf{p}}_{0,i})}. \end{aligned} \tag{7.21}$$

Assuming i.i.d. equiprobable messages, we may write

$$\Pr(\check{\mathbf{c}}_j \,|\, \check{\mathbf{p}}_{0,i}, O) = \frac{\Pr(O \,|\, \check{\mathbf{c}}_j, \check{\mathbf{p}}_{0,i})}{\Pr(O \,|\, \check{\mathbf{p}}_{0,i})} \cdot \frac{1}{2^{N_{\mathrm{S}}}}. \tag{7.22}$$

Extending the setting to the transmission of a sequence of S statistically independent messages, the probability that a specific sequence $\check{\mathbf{c}}_{j_1}, \dots, \check{\mathbf{c}}_{j_S}$ has been transmitted given the observations $O^{(1)}, \dots, O^{(S)}$ and a trial position $\check{\mathbf{p}}_{0,i}$ is the product of the individual message probabilities. Noting that the observations contain additive errors which are i.i.d. complex

Gaussian with zero mean and variance σ_{N}^2, we may write

$$\Pr\left(\check{\mathbf{c}}_{j_1},\ldots,\check{\mathbf{c}}_{j_S} \mid \check{\mathbf{p}}_{0,i}, O^{(1)},\ldots,O^{(S)}\right)$$

$$= \prod_{s=1}^{S} \Pr\left(\check{\mathbf{c}}_{j_s} \mid, \check{\mathbf{p}}_{0,i}, O^{(s)}\right)$$

$$= \frac{1}{2^{(N_{\mathrm{S}}\cdot S)}} \cdot \left(\prod_{s=1}^{S} \frac{1}{\Pr\left(O^{(s)} \mid \check{\mathbf{p}}_{0,i}\right)}\right) \cdot \left(\prod_{s=1}^{S} \Pr\left(O^{(s)} \mid \check{\mathbf{c}}_{j_s}, \check{\mathbf{p}}_{0,i}\right)\right)$$

$$= \frac{1}{2^{(N_{\mathrm{S}}\cdot S)}} \cdot \left(\prod_{s=1}^{S} \frac{1}{\Pr\left(O^{(s)} \mid \check{\mathbf{p}}_{0,i}\right)}\right) \cdot \left(\frac{1}{\pi\sigma_{\mathrm{N}}^2}\right)^{S} \cdot \exp\left(-\frac{1}{\sigma_{\mathrm{N}}^2} \sum_{s=1}^{S} \left|O^{(s)} - Z_{\mathrm{in}}^{(s)}\left(\check{\mathbf{c}}_{j_s}, \check{\mathbf{p}}_{0,i}\right)\right|^2\right).$$

$$(7.23)$$

We find the maximum likelihood estimator of the message sequence by taking the natural logarithm of (7.23) and neglecting terms independent of $\mathbf{c}_{j_s}$ as

$$\hat{\mathbf{c}}_{j_1},\ldots,\hat{\mathbf{c}}_{j_S} = \arg\min_{\check{\mathbf{c}}_{j_1},\ldots,\check{\mathbf{c}}_{j_S}} \sum_{s=1}^{S} \left|O^{(s)} - Z_{\mathrm{in}}^{(s)}\left(\check{\mathbf{c}}_{j_s}, \check{\mathbf{p}}_{0,i}\right)\right|^2. \qquad (7.24)$$

Noting that the summands can be minimized for each state s individually, we obtain

$$\hat{\mathbf{c}}_{j_s} = \arg\min_{\check{\mathbf{c}}_{j_s}} \left|O^{(s)} - Z_{\mathrm{in}}^{(s)}\left(\check{\mathbf{c}}_{j_s}, \check{\mathbf{p}}_{0,i}\right)\right|^2, \qquad (7.25)$$

justifying the independent optimization for each message in (7.17).

7.3.1 Numerical Performance Analysis

To evaluate the performance of joint decoding and localization we consider a 2D coplanar setup with one measuring anchor, N_{S} transmitters, and a single agent, all having positions chosen uniformly within a bounding box with $b = 0.5\,\mathrm{m}$. We base the optimization on the transmission of $S = 100$ messages with the noise variance chosen as $\sigma_{\mathrm{N}}^2 = 10^{-4}\,\Omega^2$. In accordance with the problem formulation presented above, we optimize the anchor position using an exhaustive search over a finite number of trial positions with $N_{\mathrm{P}} = 200 \times 200 = 40000$ grid points. Given the dimensions of the bounding box, this procedure leads to a quantization error of up to $1.8\,\mathrm{mm}$ in the position estimate, which can be resolved by introducing a subsequent gradient-based refinement of the position, as has been discussed in Chapter 6.

A primary concern lies in the question how much the localization performance is degraded by the fact that the messages $\mathbf{C}$ are not known a priori. We address this issue by investigation the

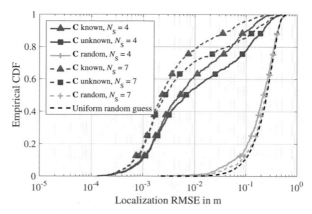

Figure 7.14. Empirical CDF of the localization RMSE for joint decoding and localization (red), compared to localization with a priori known (blue) and randomly guessed (yellow) messages.

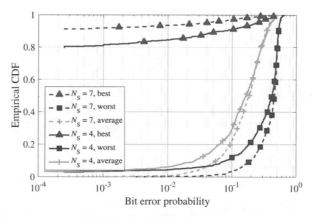

Figure 7.15. Achievable bit error probabilities for joint detection and localization.

localization performance of three different constraints: $\mathbf{C}$ being known a priori, $\mathbf{C}$ being jointly estimated with the agent position, and $\mathbf{C}$ being randomly guessed prior to the localization. We evaluate the localization RMSE over 800 random networks with 40 noise realizations each. Fig. 7.14 shows the resulting localization performance for all three cases, using networks with $N_S = 4$ and $N_S = 7$ transmitters. Herein, the solid lines and dashed lines correspond to $N_S = 4$ and $N_S = 7$, respectively, while the results for $\mathbf{C}$ being known a priori are shown in blue, and the results for joint estimation and random selection of $\mathbf{C}$ are shown in red and yellow, respectively. Furthermore the black dashed line represents a lower bound on the localization performance resulting from a blind guess of the agent position.

As can be expected, prior knowledge of $\mathbf{C}$ leads to the best performance with a sub-cm median RMSE for both settings of N_S. On the other hand, the importance of estimating $\mathbf{C}$ is clearly demonstrated by the severely degraded results of localization subject to a randomly guessed estimate of $\mathbf{C}$, for which the RMSE is close to the worst-case performance bound. Conspicuously, in this case the RMSE using 7 transmitters is worse than for the case with 4 transmitters. We conjecture that this behavior can be explained by the fact that the probability of guessing $\mathbf{C}$ correctly (or, e.g. the probability of producing only a single bit error from a random guess) is greater for $N_S = 4$ than for $N_S = 7$.

Compared to a random selection of $\mathbf{C}$, the performance loss from jointly estimating $\mathbf{C}$ with the agent position is small. Particularly for values below the median RMSE the empirical CDFs for a priori known and jointly estimated $\mathbf{C}$ show good agreement, meaning that for these realizations the transmitted messages were either estimated virtually error-free, or any estimation errors occurred only for transmitters not significant for the localization. To further investigate this point we evaluate the corresponding bit error probabilities occurring in the estimation of $\mathbf{C}$.

We evaluate the bit error probabilities for the individual transmitters of each random network, averaged over all $S = 100$ bit states and 40 noise realizations[3]. The empirical CDFs of the bit error probabilities pertaining to the best and worst transmitter of each network are shown in Fig. 7.15 in blue and red, respectively. In addition, the bit error probability averaged over all transmitters of the network is shown in yellow. As before, the solid curves represent the CDF for networks with $N_S = 4$, while the dashed lines correspond to the case of $N_S = 7$. It can be seen from these results that there is a substantial difference in the bit decoding performance among individual transmitting nodes of the random networks. The best transmitter in each network achieves a bit error probability of 1 % or better in 85 % and

---

[3] As the number of evaluated bits was equal to 4000 for each transmitter, bit error probabilities per transmitter with values below $1/4000 = 2.5 \cdot 10^{-4}$ cannot be displayed.

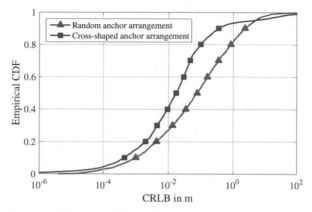

Figure 7.16. Empirical CDF of the CRLB for $N_\mathrm{S} = 4$ for random and fixed anchor arrangements with $\sigma_\mathrm{N}^2 = 10^{-4}\,\Omega^2$.

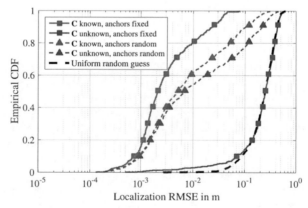

Figure 7.17. Empirical CDF of the RMSE for $N_\mathrm{S} = 4$ for random and cross-shaped anchor arrangements with $\sigma_\mathrm{N}^2 = 10^{-4}\,\Omega^2$.

93 % of the network realizations for $N_S = 4$ and $N_S = 7$ respectively. On the other hand, both the average and worst transmitter perform substantially worse, with no configuration among these achieving a bit error probability of 1 % in more than 6 % of the network realizations. Contrasting this result with the good localization performance observed in Fig. 7.14 for joint decoding and localization, we draw two conclusions. First, not all bits need to be estimated correctly in order to achieve a good estimate of the anchor location. Furthermore we conjecture that the transmitters most significant for localization are those strongly coupled to the measuring anchor, resulting in a low bit error probability for these nodes.

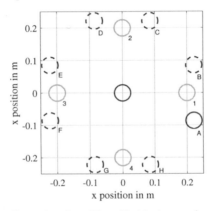

Figure 7.18. Illustration of possible ambiguities in cross-shaped topology.

In a second step we propose to improve the performance of the optimization by placing all nodes—excluding the agent—in a fixed, optimized arrangement. To this end we choose the cross-shaped topology proposed in Section 6.3.4 with the measuring anchor placed at the origin. To investigate if this topology can also beneficially be applied to single-port localization, we evaluate the CRLB for the case of the messages **C** being known a priori. These results are depicted in Fig. 7.16, which compares the CRLB evaluated over 5000 different random networks for random and fixed anchor arrangements represented by the blue and red curve, respectively. As the CRLB values of the fixed topology lie below those observed for entirely random networks in approximately 95 % of the cases, a general improvement of the localization RMSE can be expected from using the cross-shaped topology. However, an interesting effect arises when applying this network topology to joint decoding and localization, as can be observed in Fig. 7.17 which compares the localization RMSE of networks with fixed and random arrangements. We herein consider both the cases with and without a priori knowledge of the transmitted messages. When $\mathbf{C}$ is known the localization performance is improved by the fixed topology as expected. This effect is most notable for realizations with large RMSE values, with the 90th percentile of the RMSE decreasing from 10.3 cm to 2.3 cm. On the other hand, when using the cross-shaped network topology the RMSE of the joint decoding and localization is severely degraded to the point of being largely identical with the result of randomly guessing the agent position.

We conjecture that the degradation can be explained by the introduction of several minima in the cost function due to the symmetry of the anchor arrangement, as discussed in Section 7.2.1. In contrast to a setting where $\mathbf{C}$ is known, these ambiguities can not be resolved by load switching in joint decoding and localization. To illustrate this point, consider the exemplary network arrangement in Fig. 7.18. The measuring and passive anchors are depicted in red and green, respectively, and the passive anchors are numbered as indicated. The solid blue circle labeled as position A corresponds to the true agent position $\mathbf{p}_0$. It is evident that in any state s the value of $Z_{\text{in}}\left(\mathbf{p}_0, \mathbf{c}^{(s)}\right)$ is identical to the input impedance observed at positions which result from any combination of mirroring $\mathbf{p}_0$ along the symmetry axes of the anchor arrangement and rotating $\mathbf{p}_0$ by integer multiples of $\pi/2$ around the origin, as long as the bits $\mathbf{c}^{(s)}$ of the passive anchors are shifted and/or swapped accordingly. The resulting ambiguous positions, which are indicated by dashed blue lines labeled B through H, correspond to the message permutations listed in Tab. 7.1. It is important to realize that these ambiguities are

Position	Message $\mathbf{c}^{\mathrm{T}}$
A	$[c_1, c_2, c_3, c_4]$
B	$[c_1, c_4, c_3, c_2]$
C	$[c_4, c_1, c_2, c_3]$
D	$[c_2, c_1, c_4, c_3]$
E	$[c_3, c_4, c_1, c_2]$
F	$[c_3, c_2, c_1, c_4]$
G	$[c_2, c_3, c_4, c_1]$
H	$[c_4, c_3, c_2, c_1]$

Table 7.1. Cost function ambiguities and associated message permutations for cross-shaped topology.

present in all states, independent of the true value of $\mathbf{c}^{(s)}$, meaning that switching the load state of the passive anchors can not contribute to resolving them. A possible solution to this problem is to initiate the communication of the transmitters with a known training sequence to gain a rough estimate of the agent position which is sufficient to resolve the positional

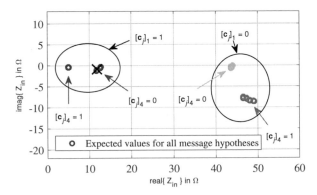

Figure 7.19. Exemplary clustering of observation space ($N_S = 7$) in the impedance domain.

ambiguity, and utilize the subsequent communication with unknown messages to refine the position estimate.

7.3.2 Reliability of Decoded Bits

As the numerical evaluations of joint decoding and localization have shown, the estimated message bits are subject to errors, the probability of which being different for each transmitter as it depends on the transmitters ability to influence the observations $O^{(s)}$ by selecting different load states. The reliability of the estimated bits pertaining to each transmitter is therefore governed by the relative arrangement of the transmitter to all other nodes in the network. To provide further intuition, Fig. 7.19 demonstrates the variable reliability of the individual transmitters in an exemplary network with $N_S = 7$ transmitters. The colored circles visualize all possible values of $Z_{\text{in}}(\mathbf{p}_0, \check{\mathbf{c}}_j)$ for the given network topology. It can be seen that the alphabet of noiseless observations falls into distinct clusters depending on the bit values of the most dominant transmitters. Particularly the bit values of transmitters 1 and 4 can be clearly distinguished. The cross in Fig. 7.19 shows a realization of a noisy observation from this network with an unknown message being transmitted, generated with a noise variance of $\sigma_N^2 = 10^{-4}\,\Omega^2$ used in the above. Based on the least-squares metric in (7.25) it is most likely that the observation belongs to a message in the blue cluster, for which $[\mathbf{c}_j]_1 = 1\,\forall j$ and $[\mathbf{c}_j]_4 = 0\,\forall j$, allowing to associate the bit values of 1 and 0 to transmitters 1 and 4, respectively, with low probability of error. The bit values of the remaining transmitters can be distinguished less easily.

To quantify the reliability of the decoded bits of the nth transmitter where $n \in \{1, \ldots, N_\mathrm{S}\}$, we define the sets

$$\mathcal{C}_n = \{j \mid [\check{\mathbf{c}}_j]_n = 1\}, \tag{7.26}$$

containing the indices of all trial messages for which the nth transmitter has the bit value 1, and its complement

$$\overline{\mathcal{C}}_n = \{j \mid [\check{\mathbf{c}}_j]_n = 0\}. \tag{7.27}$$

Note that these sets are independent of the current state index s. Given an observation $O^{(s)}$ the probability of the nth transmitter having set its bit value to 1 is given by

$$\Pr\left([\mathbf{c}]_n = 1 \mid O^{(s)}, \mathbf{p}_0\right) = \sum_{j \in \mathcal{C}_n} \Pr\left(\mathbf{c}_j \mid O^{(s)}, \mathbf{p}_0\right), \tag{7.28}$$

which depends on the agent position $\mathbf{p}_0$. Assuming i.i.d. random messages we may write

$$\Pr\left([\mathbf{c}]_n = 1 \mid O^{(s)}, \mathbf{p}_0\right) = \frac{1}{2^{N_S}} \cdot \frac{1}{\Pr\left(O^{(s)} \mid \mathbf{p}_0\right)} \cdot \sum_{j \in \mathcal{C}_n} \Pr\left(O^{(s)} \mid \mathbf{c}_j, \mathbf{p}_0\right). \tag{7.29}$$

Considering the assumed additive Gaussian noise model, we formulate the likelihood ratio

$$\Lambda(\mathbf{p}_0) = \frac{\Pr\left([\mathbf{c}]_n = 1 \mid O^{(s)}, \mathbf{p}_0\right)}{\Pr\left([\mathbf{c}]_n = 0 \mid O^{(s)}, \mathbf{p}_0\right)} = \frac{\sum_{j \in \mathcal{C}_n} \exp\left(-\frac{1}{\sigma_\mathrm{N}^2} \cdot \left|O^{(s)} - Z_\mathrm{in}^{(s)}(\check{\mathbf{c}}_j, \mathbf{p}_{0,i})\right|^2\right)}{\sum_{j \in \overline{\mathcal{C}}_n} \exp\left(-\frac{1}{\sigma_\mathrm{N}^2} \cdot \left|O^{(s)} - Z_\mathrm{in}^{(s)}(\check{\mathbf{c}}_j, \mathbf{p}_{0,i})\right|^2\right)}, \tag{7.30}$$

for which the values $\Lambda(\mathbf{p}_0) = 0$ and $\Lambda(\mathbf{p}_0) = \infty$ signify $[\mathbf{c}]_n = 1$ and $[\mathbf{c}]_n = 0$ with certainty, while $\Lambda(\mathbf{p}_0) = 1$ corresponds to maximum uncertainty. For a joint decoding and localization problem, the likelihood ratio serves as a metric to quantify the likelihood of a decoded bit from transmitter n, given the corresponding observation $O^{(s)}$ and knowledge of the agent position $\mathbf{p}_0$. If the parameters of a particular localization setup justify the assumption that the estimate of the agent position $\hat{\mathbf{p}}_0$ is generally close to its true position $\mathbf{p}_0$, the likelihood ratio can be used to perform soft estimates of the transmitted messages. To this end it is convenient to define a symmetric and bounded metric of the ratio. A sensible choice lies in the transformation $\tanh\left(\frac{1}{2}\ln\left(\Lambda(\mathbf{p}_0)\right)\right) \in [-1, 1]$. By using the identity

$$\tanh(x) = \frac{\exp(x) - \exp(-x)}{\exp(x) + \exp(-x)} \tag{7.31}$$

it can be shown that this transformation is equivalent to

$$\tanh\left(\frac{1}{2}\ln\left(\Lambda(\mathbf{p}_0)\right)\right) = \frac{\Lambda(\mathbf{p}_0) - 1}{\Lambda(\mathbf{p}_0) + 1} = 2\Pr\left([\mathbf{c}]_n = 1 \,|\, O^{(s)}, \mathbf{p}_0\right) - 1. \qquad (7.32)$$

Accordingly, values of -1 and 1 for the expression in (7.32) correspond to certainty that the bit of interest is was transmitted as 0 and 1, respectively, while the value 0 signifies maximum uncertainty.

The likelihood ratio also allows to illustrate how strongly the reliability of a particular transmitter in a given network depends on the agent position. To this end we weight the transformed likelihood ratio according to the true value of the transmitted bit, thus defining a reliability metric $\Xi(\mathbf{p}_0)$ as

$$\Xi(\mathbf{p}_0) = \begin{cases} -\tanh\left(\frac{1}{2}\ln\left(\Lambda(\mathbf{p}_0)\right)\right) : & [\mathbf{c}]_n = 0, \\ \tanh\left(\frac{1}{2}\ln\left(\Lambda(\mathbf{p}_0)\right)\right) : & [\mathbf{c}]_n = 1. \end{cases} \qquad (7.33)$$

A value of $\Xi(\mathbf{p}_0) = 1$ signifies that the bit $[\mathbf{c}]_n$ can correctly be estimated from an observation with certainty, while for $\Xi(\mathbf{p}_0) = 0$ the uncertainty about $[\mathbf{c}]_n$ is maximal. Negative values of $\Xi(\mathbf{p}_0)$ indicate that an incorrect bit estimate is more likely than a correct one, i.e. the bit error probability of $[\mathbf{c}]_n$ is greater than 0.5.

To evaluate the reliability of a specific node in a set of transmitters, we evaluate $\Xi(\mathbf{p}_0)$ for all possible agent positions. As the reliability metric is subject to the random noise contained in the observation $O^{(s)}$, we average over multiple load states of the network corresponding to observations with independent noise realizations. In Figs. 7.20(a-e), we show the resulting averaged reliability metric as a function of the agent position $\mathbf{p}_0$ for an exemplary network with $N_S = 5$. Herein the measuring anchor is drawn in red and the transmitters in green, with the transmitter of interest corresponding to the dashed circle. The values of $\Xi(\mathbf{p}_0)$ were averaged over $S = 100$ transmitted messages with independent noise realizations.

The results allow for a deeper understanding how the arrangement of the transmitters within the network influences their reliability. In Fig. 7.20a the transmitter of interest experiences a strong coupling to the measuring anchor, thus substantially influencing the observation $O^{(s)}$ and leading to a high reliability independent of the agent position. On the other hand, the transmitter being evaluated in Fig. 7.20(c) experiences a much weaker coupling to the measuring anchor. Here the reliability depends heavily on the agent position, with high reliability only being achieved for agent positions located between the respective transmitter and the measuring anchor. It can be conjectured that this effect is due to the agent acting as passive relay for the transmitter (cf. Section 4.2).

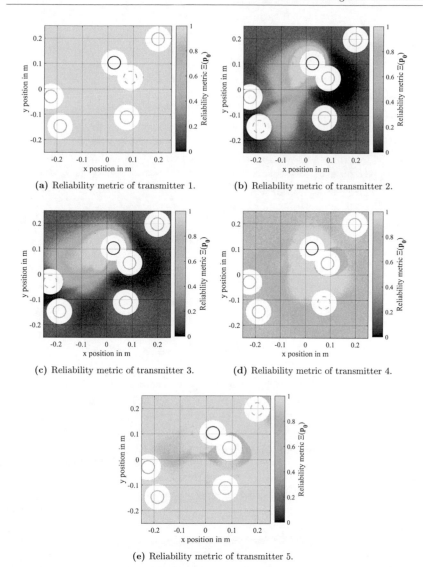

(a) Reliability metric of transmitter 1.

(b) Reliability metric of transmitter 2.

(c) Reliability metric of transmitter 3.

(d) Reliability metric of transmitter 4.

(e) Reliability metric of transmitter 5.

Figure 7.20. Reliability metric $\Xi(\mathbf{p}_0)$ of individual transmitters (dashed) in exemplary network as a function of agent position.

7.4 Conclusions

In this chapter we show that the utilization of secondary nodes with known location information allows to perform circuit-based localization with only a single measurement device. However, the network size chosen in this chapter ensures that agent and measuring anchor are typically close enough to ensure that the performed measurements are sufficiently influenced by the agent to infer its position. If the networks become larger, it may become necessary to distribute multiple measuring anchors throughout the network in order to enable reliable localization. On the other hand, passive anchors can also extend the useful range of interaction between agent and measuring anchor: in a constructed example, we demonstrate that the distance for which a sub-cm ranging RMSE is achieved can be extended by a factor of 3.

We show by simulations that a single complex impedance measurement at the anchor can suffice to locate an agent device in two dimensions, and study in which cases ambiguities may occur. By the additional introduction of load switching at the passive anchors, we can identify a synergy between localization and communication in inductively coupled networks. On one hand, we observe that load switching, as it occurs when communicating using load modulation, leads to an improvement of localization performance. This performance gain is shown to not result from noise averaging, but from a reduction of ambiguities in the underlying cost function. On the other hand we show that the bit error probability of a particular node transmitting by load modulation depends strongly on its position within the network. Obtaining the position information of a node therefore allows to predict its communication performance.

The presented localization results were based on the assumption of having a priori location knowledge for a possibly large number of passive anchors. This assumption may not be attainable in sensor networks deployed in an ad-hoc fashion. However, the idea of single-port localization assisted by passive anchors can be extended towards an iterative localization approach. Iterative localization is a multi-step process in which a subset of the agents is located in each step. Given the acquired location knowledge, these nodes assist localization of the remaining anchors in the subsequent steps [131]. In the context of our work, we envision a network containing a measuring anchor, a large number of agents and a small number of passive anchors. Iterative localization of the agents can be performed by sequentially locating a single agent, while the remaining agents disconnect their antenna to not influence the measurement. Once an agent has been located, it can act as a passive anchor in the subsequent localization step. The study of such an iterative localization approach holds interesting open questions for future work, e.g. how to reduce propagation of localization errors.

8

Practical Verification of Localization in Imperfect Environments

The investigation of circuit-based near-field localization in the previous chapters suggests a high degree of accuracy in many of the considered configurations. When judging the obtained error performance it is important to consider that the localization results were obtained by simulations using a set of simplifying assumptions. It is therefore prudent to verify the localization performance based on experimental measurements obtained from a practical implementation of the system.

This chapter will introduce an experimental localization system implementing the multi-anchor localization scheme studied in Chapter 6, and discuss its imperfections. Some of these imperfections, namely the incomplete knowledge of the manufactured circuits as well as the effect of temporal drift in the measurement instrument, can affect the observed quantities on a scale sufficient to significantly impact the outcome of the subsequent localization step. We therefore introduce two calibration mechanisms which are suited to address these imperfections while accounting for the specific requirements of the low-complexity sensor networks in which we envision circuit-based near-field localization to be primarily used. Afterwards the ranging and localization performance of the system is evaluated. Parts of this chapter have been published in [140].

8.1 Anchor and Agent Node Design

In order to experimentally perform the circuit-based localization scheme proposed in Chapter 6 we need to measure a quantity of interest, i.e. the input impedance Z_{in} or reflection coefficient

Γ_{in} for a set of physically manufactured anchors and an agent arranged with positions and orientations $\boldsymbol{\theta}_n$ and $\boldsymbol{\theta}_0$, respectively. Several anchor and agent nodes were manufactured to be employed in the measurement system. The nodes are implemented as printed circuit board (PCB) on 84 mm × 53 mm sheets of FR-4 with a substrate thickness of 1 mm and both sides being coated with a 35 μm copper sheet. Each node has a circular loop antenna with $\nu = 5$ windings and a radius of approximately $r \approx 25$ mm etched on one side of the substrate. More precisely, both the conductor width and the spacing between the individual turns are chosen as 150 μm, resulting in an inner loop radius of 23.975 mm and outer radius of 25.325 mm. The feed point of the antenna is connected to a matching network which, depending on the functionality of the device, can either be a one-port or two-port network. For the two-port case, the input port of the matching network is connected to an SMA-type connector. The circuit design used for the anchor nodes is shown in Fig. 8.1, while a photograph of a fabricated anchor is given in Fig. 8.2.

The inductor L_n and resistor R_{n1} denote the self-inductance and ohmic loss resistance of the antenna of the nth anchor. Characterization measurements of the fabricated antenna have found an inductance value of $L = 3.78\,\mu$H and an ohmic resistance of $R = 4.7\,\Omega$. The anchor matching network, consisting of the capacitors C_{n1} and C_{n2} as well as the resistor R_{n2} serves to realize a resonant match around frequency f_{res}. This frequency was chosen as $f_{\text{res}} = 25$ MHz, yielding design values of $C_{n1} = 6.23$ pF, $C_{n2} = 4.5$ pF, and $R_{n2} = 33$ kΩ for a setting in which the anchor is not loaded by the inductive coupling with other nodes. To account for this coupling as well as manufacturing tolerances of the used components, tunable capacitors were used in the implementation to enable manual adjustment of the resonance frequency to the desired value.

The circuit design of the agent node is given in Fig. 8.3. Here, m denotes the index of the agent. As the photolithographic manufacturing process of the PCB antennas has very small tolerances compared to their overall physical dimensions, we can sensibly assume $L_m = L_n$ and $R_{1n} = R_m$ for all nodes m, n. The agents are implemented as purely passive devices with the matching network being simply chosen as a single capacitor, yielding a series RLC circuit. As the resonance frequency of this network is calculated as $\omega_{\text{res}} = (LC)^{-\frac{1}{2}}$, the design value of C_m is obtained as 10.7 pF. Similar to the anchor nodes, the capacitor was chosen to be tunable to account for component tolerances, as can be seen in the fabricated agent tag shown in Fig. 8.4. It should be noted that the empty pads of the agent PCB can be used to implement load modulation and to observe induced voltages at the agent, however this functionality is not discussed within the scope of this work.

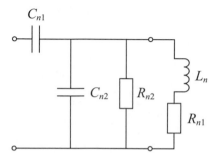

Figure 8.1. Circuit model for anchor node n.

Figure 8.2. Fabricated anchor node for experimental localization.

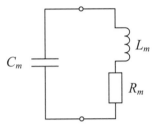

Figure 8.3. Circuit model for agent node m.

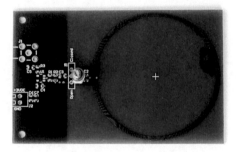

Figure 8.4. Fabricated agent node for experimental localization.

8.2   Measurement System

The presented measurement system is based on the use of an HP 8753E vector network analyzer (VNA), which allows measurement of both the input impedance $Z_{in}(\omega_w)$ as well as the reflection coefficient $\Gamma_{in}(\omega_w)$ at the discrete frequencies ω_w.

Any practical measurement instrument inherently produces measurement errors, which can generally be divided into random and systematic errors. While systematic errors can be mitigated in the measurement setup as discussed in Section 8.4, the presence of random errors is inevitable. However, for many sources of random measurement errors the perturbation is additive to the desired measurement and the underlying probability distribution function (PDF) of the error has zero mean, meaning that the error variance can be reduced by means of averaging over multiple measurements. A well-known example is thermal noise, which is modeled by additive zero-mean Gaussian noise.

Formal analysis of the noise behavior of VNAs shows that the measurements exhibit random errors from multiple sources [129], with the individual random error contributions sharing an additive zero-mean property in the S-parameter domain, although they follow different distributions. Instead of analyzing the individual error contributions, it suffices for the purposes of this analysis to characterize the total random errors by their variance σ_N^2. It should be noted that these considerations hold only for measurements of the reflection coefficient, which are directly obtained by the VNA in the S-parameter domain due to its operating principle [20], while the input impedance measurements are obtained by converting the noisy reflection measurements using relation (6.7), thus transforming the noise.

To assess the noise level of the measurement instrument we perform $N = 300$ measurement sweeps of both the reflection coefficient and the input impedance. Each sweep measures the quantity of interest at $W = 401$ frequency samples ω_w in the range of 24 MHz to 26 MHz using an intermediate frequency (IF) filter bandwidth of the VNA set to 1000 Hz. The choice of IF filter bandwidth has an immediate impact on the noise level of the measurement. Choosing a filter with a smaller bandwidth reduces the noise contributing to each measurement, but increases the acquisition time for all L measurements. The total acquisition time of the 401 measurements is about 0.54 s using the setting of 1000 Hz. For the example of the reflection coefficient, an estimate of the noise variance at frequency ω_w is then obtained by

$$\hat{\sigma}_N^2(\omega_w) = \frac{1}{N-1} \sum_{n=1}^{N} \left| \Gamma_n(\omega_w) - \left(\frac{1}{N} \sum_{m=1}^{N} \Gamma_m(\omega_w) \right) \right|^2 , \tag{8.1}$$

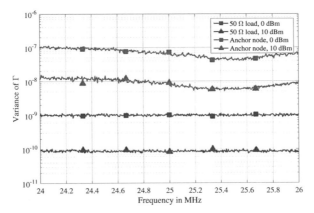

Figure 8.5. Measured variance estimate of the input reflection coefficient Γ for different input power levels of the VNA.

where the term $\frac{1}{N-1}$, referred to as Bessel's correction, accounts for the bias in the variance estimate.

We characterize the noise variance using both a standard $50\,\Omega$ termination as well as one of the manufactured anchor nodes as load for the VNA. The measurement variance of the reflection coefficient is shown in Fig. 8.5.

We have used two different settings for the input power of the VNA, namely the default setting of $0\,$dBm as well as an increased input power of $10\,$dBm. It can be clearly seen that increasing the input power results in a lower noise variance. However, choosing the input power too high can introduce another source of measurement errors known as compression errors, which stem from the analog components of the VNA being driven into a nonlinear operating regime. We restrict ourselves to a maximum input power of $10\,$dBm, as significantly higher power settings can produce an overload of the VNA measurement port.

As can be expected, the noise variance measured at the $50\,\Omega$ termination is invariant in frequency. The anchor circuit, on the other hand, has a resonant behavior that is also reflected in the noise variance measurement. Particularly in the input impedance domain, the resonance at $25.4\,$MHz is clearly visible.

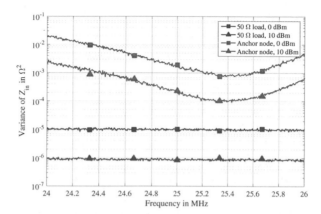

Figure 8.6. Measured variance estimate of the input impedance Z_{in} for different input power levels of the VNA.

8.2.1   Localization Based on Virtual Anchors

A straightforward measurement approach to perform experimental localization for a network specified by anchor parameters $\boldsymbol{\theta}_n$ and an agent parameters $\boldsymbol{\theta}_0$ is to arrange the nodes according to their respective positions and orientations and connect the measurement port of the VNA either sequentially if only a single measurement instrument can be used, or simultaneously given that multiple VNAs are available. As discussed in Section 6.2 we utilize the assumption of each measurement only depending on the interaction of a single anchor-agent pair. This can easily be achieved by sequential measurement at the anchors with one anchor measuring while the remaining antenna circuits are disconnected. In consequence the non-measuring anchors do not interact with the current measurement and can be disregarded. Alternatively it is possible to reformulate the optimization process such that the effect of inter-anchor coupling is taken into account in the system model for the input impedance or reflection coefficient.

It is desirable to investigate the experimental localization performance for a large number of different anchor-agent networks, thus enabling to gauge the impact of both random network geometries as well as variable properties such as the number of anchors. However, the key drawback of the above described measurement setup lies in the fact that manually arranging the localization network to a given set of $\boldsymbol{\theta}_n$ and $\boldsymbol{\theta}_0$ is a tedious task. More importantly, the true positions and orientations of both the anchors as well as the agent need to be known: the anchor parameters are required in order to perform the optimization, and the agent parameters are needed to calculate the localization error. Manual node placement is imprecise and would introduce additional errors in the optimization process.

To efficiently automate the measurement of several network arrangements as well as to guarantee precise and repeatable node placement, we introduce the concept of a virtualized anchor setup. This concept exploits the approach of sequential measurements at all anchors, resulting in no inter-anchor coupling. The goal of using virtual anchors is to obtain the measurements corresponding to a particular *virtual* network with N_A defined by $\boldsymbol{\theta}_n$, $\boldsymbol{\theta}_0$, while using only a single physical anchor-agent pair. To this end it is assumed that the circuit elements of all virtual anchors in the network are identical to those of the physical anchor.

The physical anchor node is statically installed on the measurement port of the VNA. For each virtual anchor in the sequence of measurements, the physical anchor's position and orientation are assumed to be identical to the values of $\boldsymbol{\theta}_n$ in the coordinate system of the virtual network topology. The physical agent, on the other hand, can be arranged in space such that the relative position and orientation of the physical anchor-agent pair is identical to the relative position and orientation of the respective virtual anchor-agent pair. In effect, the expected value of each measurement is identical for the physical and virtual pair.

After measurement, the agent node is displaced to emulate the relative arrangement of the next virtual anchor-agent pair in the measurement sequence. This displacement can be performed in a precise and automated fashion by using a positioner device. In the presented measurement setup the physical agent is mounted on the arm of a HIGH-Z S-1000 three-axis positioner with an operational range of $1000\,\text{mm} \times 600\,\text{mm} \times 110\,\text{mm}$. The positional accuracy of the device is specified as $30\,\mu\text{m}$ with a minimum step size of $1.8\,\mu\text{m}$. As the employed positioning device does not allow for manipulation of the orientation of the agent, the realization of arbitrary mutual orientations is restricted with this setup. The agent has therefore been mounted in a fixed orientation which is chosen identical to the physical anchor as $\mathbf{q} = \mathbf{0}$, allowing the measurement system to generate measurements for 1D, 2D, and 3D network topologies. Fig. 8.7 shows a photograph of both the physical anchor and physical agent in the measurement setup.

8.3   Circuit Calibration

We chose the frequency-dependent reflection coefficients $\Gamma_{\text{in},n}(\omega_w)$ as measurement quantities to base the localization on. For notational convenience, the entire set of reflection coefficients for all anchors and frequencies is expressed as matrix $\boldsymbol{\Gamma}_{\text{in}}$ with rows representing the

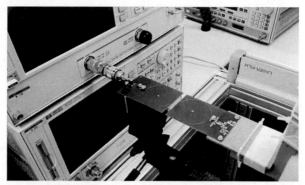

Figure 8.7. Photograph of the virtualized measurement setup. The anchor (on the left) is connected to a VNA, while the agent (right) can be moved with a positioner.

anchor index n and columns denoting the frequency index w. In an experimental setup, the localization problem in (6.13) can be formulated as

$$\{\hat{\mathbf{p}}_0, \hat{\mathbf{q}}_0\} = \arg\min_{\check{\boldsymbol{\theta}}_0} \left\| \mathbf{O}_n(\boldsymbol{\theta}_0) - \boldsymbol{\Gamma}_{\mathrm{mod}}(\check{\boldsymbol{\theta}}_0) \right\|_{\mathrm{F}}^2 . \tag{8.2}$$

Here $\mathbf{O}_n(\boldsymbol{\theta}_0)$ is a matrix containing the noisy observations of Γ_{in} at their respective anchors and frequencies, while $\boldsymbol{\Gamma}_{\mathrm{mod}}(\check{\boldsymbol{\theta}}_0)$ is a model predicting the true values Γ_{in} for each trial value $\check{\boldsymbol{\theta}}_0$. It is important to differentiate between the true values and the model because—in contrast to the assumption of perfect circuit knowledge made in previous chapters—the true value of practically measured input reflections can only be predicted from circuit theory and electromagnetics with limited accuracy, i.e. $\boldsymbol{\Gamma}_{\mathrm{mod}}(\check{\boldsymbol{\theta}}_0)$ is not necessarily identical to $\Gamma_{\mathrm{in}}(\check{\boldsymbol{\theta}}_0)$. This limitation stems from the fact that even though the circuit layout of both anchor and agent nodes is likely to be known, the individual circuit components can significantly deviate from their specification due to manufacturing tolerances, aging, and temperature dependency. Additionally, parasitic elements in the circuits are often not known or only roughly characterized. As a result, the localization approach discussed in Chapter 6 can fail, because the optimization in (8.2) does not account for the fact that the difference between an assumed reflection model and the true reflection can be significantly greater than the actual change in reflection due to the agent position and orientation.

To solve the problem of incomplete circuit knowledge, we introduce a *circuit calibration* step before the localization process: by recording a database that maps all possible relative orientations between anchor and agent to corresponding observed reflection values $\boldsymbol{\Gamma}_{\mathrm{cal}}$, the cost function in (8.2) can be evaluated by replacing the analytical expression of $\mathbf{\Gamma}_{\mathrm{mod}}(\breve{\boldsymbol{\theta}}_0)$ by a simple database lookup. Circuit calibration thus allows to perform localization even when the involved circuits of anchors and agent are completely unknown.

Recording such a database for the entire parameter space of the agent is a tedious process. However, the necessary number of measurements can be significantly reduced by further analysis of the problem. As discussed in the previous chapters, the only circuit component that depends on the unknown parameters of the agent is the mutual inductance M_n. It can be in turn expressed in terms of the self-inductances of the agent and anchor as $M_n = k_n\sqrt{L_0 L_n}$. The coupling coefficient k_n of the anchor-agent pair can take on values $-1 \leq k_n \leq 1$. We may therefore conclude that agent arrangements resulting in the same coupling coefficient k_n will produce identical values $\Gamma_{\mathrm{in},n}$. In effect, we can parametrize the reflection values to be recorded by the coupling coefficient k_n, i.e. we obtain a database of all possible observations $\Gamma_{\mathrm{in},n}(\boldsymbol{\theta}_0)$ by recording a one-time measurement of $\Gamma_{\mathrm{in},n}$ over a set of positions and/or orientations which correspond to all practical values of k_n. An example of such a set of measurement points is a coaxial arrangement of anchor and agent, with the agent increasingly being moved further away. The reflection values between these measured samples may later be inferred by interpolation.

Using a one-dimensional calibration database, the true value of $\Gamma_{\mathrm{in},n}(\breve{\boldsymbol{\theta}}_0, \omega_w)$ for any trial arrangement $\breve{\boldsymbol{\theta}}_0$ is equivalent to the recorded value $\Gamma_{\mathrm{in},n}(k_n(\breve{\boldsymbol{\theta}}_0), \omega_w)$, where $k_n(\breve{\boldsymbol{\theta}}_0)$ is the coupling coefficient corresponding to the trial arrangement. The primary benefit of this procedure is that obtaining a one-dimensional calibration measurement can be performed significantly faster and with a much smaller memory requirement than recording $\Gamma_{\mathrm{in},n}(\boldsymbol{\theta}_0, \omega_w)$ for every possible value of $\boldsymbol{\theta}_0$. Using one-dimensional calibration, the localization problem in (8.2) is reformulated as

$$\{\hat{\mathbf{p}}_0, \hat{\mathbf{q}}_0\} = \arg\min_{\breve{\boldsymbol{\theta}}_0} \left\| \mathbf{O}(\boldsymbol{\theta}_0) - \mathbf{\Gamma}_{\mathrm{cal}}(\mathbf{k}(\breve{\boldsymbol{\theta}}_0)) \right\|_{\mathrm{F}}^2, \tag{8.3}$$

where $\mathbf{\Gamma}_{\mathrm{cal}}$ represents the calibration database, and the vector $\mathbf{k}$ contains the coupling values for all anchors:

$$\mathbf{k}(\breve{\boldsymbol{\theta}}_0) = \left[k_1(\breve{\boldsymbol{\theta}}_0), \ldots, k_N(\breve{\boldsymbol{\theta}}_0) \right]^{\mathrm{T}}. \tag{8.4}$$

For this optimization procedure, the mapping function $\boldsymbol{\theta}_0 \mapsto k_n(\boldsymbol{\theta}_0)$ must be known, which is only dependent on the geometry of the involved antennas and their relative arrangement. In the 2D coplanar arrangement with circular loop antennas considered in this chapter, the coupling coefficient is found as a function of only the relative distance between respective anchor and agent. For a general six-dimensional arrangement, this mapping can be calculated from the Neumann integral (cf. (3.15)) or the dipole approximation (cf. (3.20)) of the mutual inductance in conjunction with knowledge of the self-inductances of the antennas.

8.4 Drift Autocalibration

Another source of measurement imperfections is a slowly changing systematic deviation known as drift error, which is produced by the measurement instrument. Drift is caused by changes in the environment (e.g. temperature or humidity) and the internal state of the measurement instrument. A set of circuit calibration measurements Γ_{cal} will therefore only predict measured reflection coefficients with high accuracy immediately after recording the database. To illustrate the drift errors, Fig. 8.8 shows the measured values of the real part of Γ_{in} for a coplanar node arrangement such as the one in Fig. 8.7. For both the red and blue curves, the input

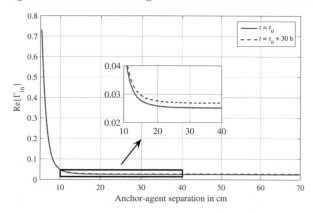

Figure 8.8. Illustration of drift error. Both lines represent measurement sweeps along identical arrangement and at the same frequency, but 30 hours apart.

reflection was measured at 25.4 MHz along a set of identical positions with increasing anchor-agent separation[1] and fixed orientations. The two measurement sweeps were conducted in a laboratory environment with a 30 h time span between them. While both measurements are similar, a close-up investigation of both curves shows a systematic deviation between them which is of sufficient magnitude to impair the mapping from measurement to anchor-agent

---

[1] The anchor-agent separation is defined as the distance between the centers of both loop antennas.

separation in this example, and thus the localization performance in general. To rule out external interference as cause of the systematic deviation, the measurements were repeated in an electromagnetically shielded environment, where the same effect was observed.

A set of models have been developed in literature to describe systematic errors in network analyzers. For an overview of these models and the physical origins of the errors, the reader is referred to [128]. Therein, the models have in common that they simplify to a description by three error terms in the case of one-port measurements, modeling the systematic drift error as an additional, unknown error two-port between the device under test (DUT, in this case the anchor circuit) and an error-free measurement instrument, as indicated in Fig. 8.9. Here

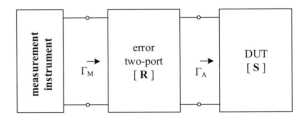

Figure 8.9. Error model for one-port reflection measurement.

Γ_A is the true input reflection coefficient, Γ_M denotes the erroneous measured value subject to drift, and

$$\mathbf{R} = \begin{bmatrix} r_{11} & r_{12} \\ r_{21} & r_{22} \end{bmatrix} \tag{8.5}$$

describes the error two-port in the form of its scattering parameter matrix, which is generally frequency dependent. With knowledge of the elements of $\mathbf{R}$, the true reflection coefficient can be calculated using the relation

$$\Gamma_A = \frac{\Gamma_M - r_{11}}{r_{22}(\Gamma_M - r_{11}) - r_{12}r_{21}}. \tag{8.6}$$

To obtain the three error terms r_{11}, r_{22}, and $r_{12}r_{21}$ several one-port calibration mechanisms have been developed. These mechanisms are generally based on the measurement of reference standards which are connected to the measurement instrument instead of the agent (equivalent to the characterization of reciprocal two-ports using a single measurement port, as discussed in Section 2.2.1). The circuit parameters of the standards often need to be fully known, e.g. for the often-used open-short-match technique [56] or can be partially unknown using so called self-calibration techniques [40].

In a sensor network setting, these calibration procedures are not practical, especially as they need to be carried out regularly to maintain adequate localization capability. On the other hand, we may utilize the insight that the circuit-based localization scheme described in Chapter 6 and its underlying cost function (6.13) are invariant to a systematic error in the observation, as long as the error is identical for the observation and the system model. It is therefore only necessary to perform a relative calibration of the measurements.

We accordingly introduce an alternative means of obtaining an estimate of the error matrix, which we refer to as *drift autocalibration*. To this end we assume a sequential interaction of anchors and agent as an the previous section. We furthermore define the initial circuit calibration measurements, which take the role of the system model in the localization cost function, to be reference values free of systematic drift errors.

The goal of drift autocalibration is to obtain estimates $\hat{\mathbf{R}}_n(\omega_w)$ of the error matrix for each anchor n and frequency point ω_w at any given time after the initial circuit calibration. The basis for obtaining these estimates is the insight that the anchors will retain the true value of their input reflection coefficient given a specific coupling, i.e. the relation

$$\Gamma_{\mathrm{A},n}(k_n, t_1) = \Gamma_{\mathrm{A},n}(\check{k}_n, t_2) \tag{8.7}$$

holds true if $\check{k}_n = k_n$, independent of the measurement times t_1 and t_2. It is therefore possible to use measurements at different points in time for which the coupling coefficient is known and identical to estimate the relative drift error between the measurement instants.

If the sensor network has the ability to move the agent to specific locations during drift autocalibration, the implementation of the drift autocalibration process estimates $\mathbf{R}_n(\omega_w)$ from measurements which impose the same coupling as measurements previously performed for the initial circuit calibration database Γ_{cal}. To this end, we define a set of P autocalibration points, $\mathcal{P} = \{\boldsymbol{\theta}_1, \ldots, \boldsymbol{\theta}_S\}$. Here $\mathcal{P}$ may be any subset of the measurement points used for Γ_{cal}. During the drift autocalibration process of anchor n, we obtain P measurements of the input reflection coefficient for each frequency point ω_w, denoted as $\Gamma_{\mathrm{M},n}(\omega_w, \boldsymbol{\theta}_p)$. Each of these measurements corresponds to a value $\Gamma_{\mathrm{cal},n}(\omega_w, \boldsymbol{\theta}_p)$ recorded for the same arrangement. For any trial realization of the error matrix $\check{\mathbf{R}}_n(\omega_w)$, we can calculate an estimate of the true input reflection coefficient $\Gamma_{\mathrm{A},n}(\omega_w, \boldsymbol{\theta}_p)$ using (8.6). This value should ideally be identical to $\Gamma_{\mathrm{cal},n}(\omega_w, \boldsymbol{\theta}_p)$. We therefore optimize $\hat{\mathbf{R}}_n(\omega_w)$ such that the squared error between the calibration database and autocalibration measurements is minimized:

$$\hat{\mathbf{R}}_n(\omega_w) = \underset{\hat{\mathbf{R}}_n(\omega_w)}{\arg\min} \sum_{\boldsymbol{\theta}_p \in \mathcal{P}} \left| \Gamma_{\mathrm{A},n}(\omega_w, \boldsymbol{\theta}_p) - \Gamma_{\mathrm{cal},n}(\omega_w, \boldsymbol{\theta}_p) \right|^2 . \tag{8.8}$$

We implicitly make two assumptions. On one hand we require that the values $\Gamma_{\mathrm{A},n}(\omega_w, \boldsymbol{\theta}_p)$ sufficiently differ over the set of autocalibration points, which would otherwise lead to an underdetermined problem. On the other hand, we assume that the set of reference values in the calibration database as well as the set of autocalibration measurement values were recorded in a short time period with respect to the temporal change of $\mathbf{R}_n(\omega_w)$, such that a single trial of the error matrix is applied for all contributions in (8.8).

8.4.1 Open-Circuit Autocalibration

Performing measurements at specific locations during drift autocalibration requires the ability to move the agent to these locations, which is an unrealistic constraint for many sensor network applications. Instead, the set of autocalibration points $\mathcal{P} = \{\boldsymbol{\theta}^{(\infty)}\}$ may be used, which represents a single measurement for $k = 0$, i.e. for an infinitely far away agent, which can be practically achieved by using a switch to open circuit the agent's antenna.

For the single-sample case ($P = 1$), the drift autocalibration reduces to

$$\hat{\mathbf{R}}_n(\omega_w) = \underset{\hat{\mathbf{R}}_n(\omega_w)}{\arg\min} \left| \Gamma_{\mathrm{A},n}\left(\omega_w, \boldsymbol{\theta}^{(\infty)}\right) - \Gamma_{\mathrm{cal},n}\left(\omega_w, \boldsymbol{\theta}^{(\infty)}\right) \right|^2 , \tag{8.9}$$

which is an underdetermined problem, thus possibly leading to incorrect estimates of $\mathbf{R}_n(\omega_w)$. It is possible to mitigate this issue by using a sliding window approach in frequency domain. To this end, the cost function in (8.9) can be extended to include not only the currently considered frequency, but also the Q measurements carried out for a number of neighboring frequencies:

$$\hat{\mathbf{R}}_n(\omega_w) = \underset{\hat{\mathbf{R}}_n(\omega_w)}{\arg\min} \sum_{m=w-Q/2}^{w+Q/2} \left| \Gamma_{\mathrm{A},n}\left(\omega_m, \boldsymbol{\theta}^{(\infty)}\right) - \Gamma_{\mathrm{cal},n}\left(\omega_m, \boldsymbol{\theta}^{(\infty)}\right) \right|^2 . \tag{8.10}$$

In applying such a window for the minimization at each frequency point, we implicitly assume that the error matrix changes only slowly over frequency and neighboring frequency points thus contribute additional information to the minimization problem. The benefit of using this procedure is illustrated in Fig. 8.10. This graph compares the mean value of the squared

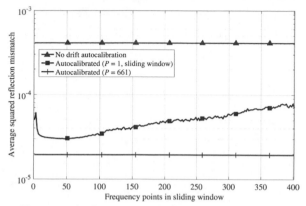

Figure 8.10. Mean squared reflection mismatch after different modes of drift autocalibration.

reflection mismatch $|\Gamma_A - \Gamma_{cal}|^2$, after no drift autocalibration, single-sample autocalibration with $P = 1$, and a full autocalibration with $P = 661$. The plotted values were averaged over 401 frequency samples and 19 autocalibration measurements, recorded over a period of 30 hours to allow for significant drift. For the single-sample calibration a sliding window approach was used, where the window size, i.e. the number of neighboring samples included in the optimization, was varied. An optimum in the reflection mismatch is observable for window sizes around 50. These results show that, compared to the mismatch without drift autocalibration, even a single autocalibration sample with an open-circuit agent achieves a substantial improvement. This can also be seen from Fig. 8.11, which once again compares the input reflection of a single anchor-agent pair for a coaxial distance sweep. In this cutout of the measurement results already discussed in Fig. 8.8, the added yellow and purple curves represent the values of the later measurement sweep, recorded at $t_0 + 30\,\mathrm{h}$, when subjected to a full autocalibration and open-circuit autocalibration, respectively. In this example, the autocalibrated measurements and their reference values show a very good match.

8.5 Measurement Results

For the evaluation of measurement-based localization, we employ the same setup introduced in Section 6.3: a two-dimensional coplanar node arrangement with one agent and $N_A = 5$ anchors, i.e. the node positions are given by $\mathbf{p}_0 = [x_0, y_0]^T$ for the anchor and $\mathbf{p}_n = [x_n, y_n]^T$ for the agents. The agent's position is randomly chosen from a uniform distribution inside

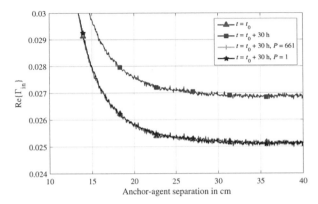

Figure 8.11. Measurements of input reflection with 30 h time difference with and without application of drift autocalibration.

a square bounding box with a side length of $b = 50$ cm. The anchors may either also be randomly placed within the bounds or fixed in the optimized cross-shaped topology discussed in Section 6.3.4.

The tunable capacitors of the anchor and agent nodes were manually adjusted to achieve resonance for close proximity at 25.4 MHz. The anchor's input reflection $\Gamma_{in,n}(f)$ is recorded at 401 linearly spaced frequency points between 25.0 MHz and 25.7 MHz with a low IF bandwidth of 1 kHz.

To perform localization within the virtualized setup, the input reflections Γ_{in} were measured for a set of coplanar anchor-agent arrangements, ranging from the closest possible distance of 53 mm to the maximal separation allowed by the bounding box, $\frac{\sqrt{2}}{2}$ m, in steps of 1 mm. This measurement sweep was repeated 100 times. A close-up of one of the recorded sweeps is displayed in Fig. 8.12 for distances up to 20 cm.

For each virtualized network, 50 randomly selected distance sweeps are used to create the initial calibration database Γ_{cal} by performing drift autocalibration—with the first sweep in the database as reference—to eliminate any drift that may have occurred during the calibration process. Subsequently the calibration database is averaged across repetitions to reduce noise. The 50 remaining measurement sweeps contain noise realizations which are independent of those in the calibration database. They are therefore used to generate the measurements for localization. To this end a subset containing M of the remaining measurement sweeps is randomly selected and averaged over the noise realizations. We initially set M to the maximum

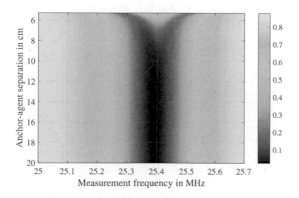

(a) Real part.

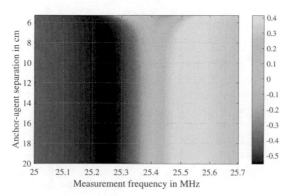

(b) Imaginary part.

Figure 8.12. Constituent measurement sweep for circuit calibration database. The real and imaginary parts of Γ_{in} are measured over distance and frequency.

value of 50. To generate the observation matrix $\mathbf{O}(\mathbf{p}_0)$ for continuous positions $\mathbf{p}_0$ from the measurement sweeps taken at discrete distances, we linearly interpolate between the noise-averaged observations at the two points neighboring the true distance of $\mathbf{p}_0$ to the anchor of interest. To increase performance of the subsequent localization, drift autocalibration may be performed on the constituent measurement sweeps before averaging across noise. In this case, the first measurement sweep of the calibration database $\mathbf{\Gamma}_{\mathrm{cal}}$ is again used as reference.

8.5.1 Experimental Ranging Performance

We first assess the usable coupling range of this measurement setup by performing localization for a network with a single anchor, i.e. distance estimation, by exhaustively searching for the minimum of the cost function in (8.3), evaluated over the entire set of possible distances with a step size of $5\,\mu\mathrm{m}$. The resulting root mean squared error (RMSE) of the distance estimates is shown in Fig. 8.13 for no drift compensation of the measured data, as well as open-circuit ($P = 1$) and full ($P = 661$) autocalibration. The results show that the range estimate is very accurate at separations below 10 cm even for the case without calibration (shown in red). In this region the oscillating behavior of the RMSE is a consequence of the evaluation of the cost function at discrete distances which introduces a quantization error of up to $2.5\,\mu\mathrm{m}$.

At greater distances, the ranging performance degrades quickly due to the near-field nature of inductive coupling. The cause of the observed flattening is discussed in Section 6.3 based on simulations. Furthermore the impact of measurement drift can clearly be seen. The maximum usable range of the used setup significantly increases to on the order of 30 cm if drift autocalibration is employed. For the blue curve, 661 agent positions of the circuit calibration database were used as reference, while the green curve corresponds to the single-sample set of $\mathcal{P} = \{\mathbf{p}^{(\infty)}\}$, i.e. only the open circuit measurement was used for drift autocalibration. To this end a sliding window approach with the previously determined optimum size of 50 samples was employed. While the open-circuit autocalibration performs suboptimal, its achieved improvement of the ranging process is comparable to the full calibration case.

8.5.2 Experimental Localization Performance

The two-dimensional localization performance was then assessed for 6000 virtual network realizations with $N = 5$ anchors. For each of these networks localization was performed using the two-step optimization procedure described in Chapter 6. Herein, an initial position estimate is first obtained using a grid search to minimize (8.3) in the vicinity of the anchor

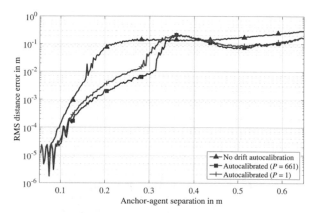

Figure 8.13. RMSE for distance estimation.

with the strongest coupling. This initial estimate is subsequently refined using the Nelder-Mead simplex algorithm[2]. The empirical CDF of the resulting localization error is shown in Fig. 8.14. The red curve describes the localization performance for randomized anchor positions and without drift compensation. A median error of 1.35 cm is achieved in this case.

The blue curves in Fig. 8.14 show the localization error of the same setup, where drift autocalibration of the measurements was additionally performed for $P = 661$. This leads to a significant improvement of almost an order of magnitude for the median error, leading to a value of 1.9 mm. On the other hand, the dashed blue line corresponds to measurements with open-circuit autocalibration. With a median error of 2.7 mm there is again a performance loss compared to the full autocalibration. However, the improvement over the case with no auto-calibration is still substantial. In an effort to further improve the localization performance, we have repeated the measurements for a fixed, heuristically optimized anchor geometry where one anchor was placed in the center of the bounding box, and the remaining four were distributed close to the center of each boundary edge. The optimization of the anchor topology is discussed in Section 6.3. The curves pertaining to the localization error of this configuration with drift autocalibration are shown in green. It can be seen that an additional gain can be achieved by carefully choosing the anchor positions, which yields a median error of 1.1 mm for the full autocalibration and 1.7 mm for single-sample autocalibration.

---

[2]It should be noted that the estimate of the refinement step was not constrained by the bounding box, allowing localization errors greater than $\frac{\sqrt{2}}{2}$ m.

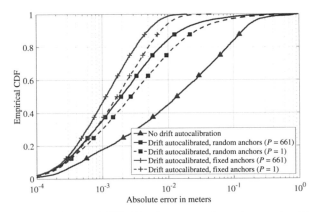

Figure 8.14. Empirical CDF of localization error for random agent position.

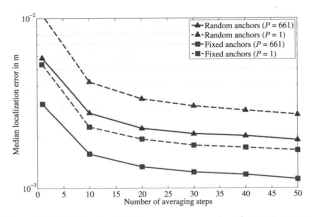

Figure 8.15. Median localization error versus number of averaging steps.

To evaluate the dependency of the localization error on the noise level present in the measurement, we vary the number of noise realizations included to generate the reflection measurements, as the number of steps is inversely proportional to the noise variance. Fig. 8.15 shows the resulting median localization error for the measurement averaging being varied from 1 to 50 steps, while the circuit calibration database was still performed with 50 averaging steps. The red curves once again correspond to the case of a random anchor topology, while green represents the results for the fixed anchor layout. Above an averaging factor of 20 the error improves only marginally, suggesting a suitable tradeoff between acquisition time and localization performance for this value.

As final result, we investigate the impact of the network's anchor density on the localization performance. To this end, the size of the bounding box is kept constant, while the number of anchors is increased from a minimum number of 3 to a total of 15 anchors. Fig. 8.16 shows the empirical CDF of the localization error for the exemplary case of random anchor arrangement and single-sample drift autocalibration. It can be seen that localization error

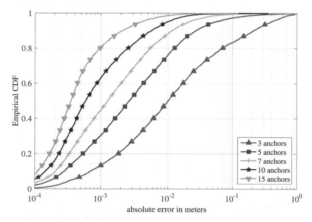

Figure 8.16. Empirical CDF of the localization error for random anchor arrangement and single-sample drift autocalibration.

performance drastically improves with anchor density, with more than an order of magnitude between the median errors of the two extreme cases. This can be explained by the fact that coupling coefficient between anchor and agent decreases fast over distance, and a dense network therefore increases the probability of obtaining a sufficient number of impedance measurements large agent contributions.

8.6   Conclusions

We experimentally demonstrate the feasibility of applying circuit based near-field localization in inductively coupled sensor networks. In contrast to simulations which may assume perfect knowledge, the presented system must deal with imperfections encountered in a practical measurement setup. This is achieved by introducing two calibration steps. On one hand, we employ an initial circuit calibration which removes the need for knowing the circuits at anchor and agent node. This step also implicitly removes the impact of any systematic errors present in the measurement setup, as they are regarded as part of the circuit. On the other hand, drift autocalibration is introduced as efficient means of compensating temporal drift errors of the measurement device without the need for connecting calibration standards. The obtained results show that the localization error is strongly dependent on the quality of the performed calibration.

Part IV.

Outlook and Conclusions

9
Summary and Conclusions

In this thesis we have investigated the viability of microsensor networks employing inductive coupling and cooperative relay devices. Building on a system model derived from circuit theory, we studied the application of inductive coupling as a physical layer for both communication and localization. In the presumed setting of dense microsensor networks consisting of many devices we intended to exploit the availability of idle, secondary nodes. In both areas we came to the conclusion that these secondary devices—by assisting the communication and localization as relays or passive anchors—can enable, facilitate, or simplify the task at hand. We state our key results as follows.

Reliable communication in inductively coupled microsensor networks can be achieved by passive relaying, even if the network is randomly arranged. We address a key limitation of inductively coupled communication: strong path loss as a result of destinations being either far away from or misaligned to the transmitter. To this end we analyze passive relaying based on the magnetoinductive waveguide effect. This method presents a suitable relaying scheme for the use in microsensor networks due to its extremely low complexity requirements for the relays, but has previously only been investigated for regular relay geometries. To investigate random relay placements, such as would be found in a practical sensor network setup, we introduce a system model which allows the analysis of arbitrary network arrangements, investigate the achievable SNR gain, and demonstrate the necessity for optimization of the relay load impedances. The presence of passive relay nodes is found to enable reliable communication over distances or for relative antenna alignments that would otherwise be prohibitive. Furthermore, we verify the practical applicability of passive relaying in an experimental setup.

Cooperative relaying in microsensor networks enables distributed computation. We introduce a novel a scheme in which relay nodes enable over-the-air computation in networks consisting of low-complexity microsensor devices. Herein the network wirelessly implements an artificial neural network of the multilayer feedforward form. The scheme is based on the ability of a sufficiently large set of half-duplex relays to arbitrarily shape a MIMO communication channel by appropriately selecting the relay gains. We show that the relay gain amplitudes which implement a desired ANN functionality can directly be found given knowledge of global channel state information. Particularly in large microsensor networks, this requirement can be prohibitive. As an alternative we introduce methods to find the relay gains in a decentralized, iterative fashion for which the relays only require knowledge of their local CSI as well as feedback terms that do not scale with the number of relays. We investigate the robustness of the proposed wireless ANN architecture to the imperfections encountered in a wireless microsensor environment, finding that the tolerance to imperfections generally increases as the number of employed relays grows.

Accurate localization of completely passive agent nodes is possible in inductively coupled systems. Using passive relay devices as secondary anchors, localization is enabled with only a single measuring anchor. We have presented a novel near-field localization scheme that allows to estimate the position of a purely passive sensor device by means of measurement of network parameters such as input impedance or reflection coefficients at multiple anchors. Analysis of this scheme has been performed by derivation of theoretical limits, simulations, and experimental measurements, all of which achieved accurate localization results. For the latter case low-complexity calibration mechanisms suitable for microsensor networks have been proposed to mitigate the imperfections of a practical measurement setup. We showed that the proposed localization scheme can be reduced to only a single measuring anchor by employing secondary nodes as passive, i.e. non-measuring anchors which reduce the ambiguity in the underlying optimization problem by their presence. The reduction of ambiguity can further be enhanced if the passive anchors perform a synchronized switching of their antenna loads, even if the sequence is unknown at the measuring anchor. The proposed localization scheme is not only suitable for low-complexity sensor-network settings, but also can be employed for accurate tracking of RFID and NFC devices.

A

Derivation of Cramér-Rao Lower Bound for Single-Port Localization

We extend the Cramér-Rao bound on the localization error for circuit-based localization, as derived in Section 6.2.4, to obtain a bound on the localization error of the single anchor localization discussed in Chapter 7. To this end we assume a 2D coplanar setup with passive anchors having known loads. Given the observations $O_w^{(s)}$ at frequency indices w and load states s, the Fisher information matrix is obtained as [165]

$$\mathbf{F}_{\mathbf{p}_0} = \mathsf{E} \left[\sum_{s=1}^{S} \sum_{w=1}^{N_{\text{freq}}} \left(\frac{\partial \ln \Pr\left(O_w^{(s)} \mid \mathbf{p}_0\right)}{\partial \mathbf{p}_0} \right) \left(\frac{\partial \ln \Pr\left(O_w^{(s)} \mid \mathbf{p}_0\right)}{\partial \mathbf{p}_0} \right)^{\mathrm{T}} \right]. \quad (A.1)$$

The log-likelihood function of the observations $\ln \Pr\left(O_w^{(s)} \mid \mathbf{p}_0\right)$ follows from the assumption of additive circularly symmetric complex Gaussian noise as

$$\ln \Pr\left(O_w^{(s)} \mid \mathbf{p}_0\right) = \ln\left(\frac{1}{\pi \sigma_{\mathrm{N}}^2}\right) - \frac{\left| O_w^{(s)} - Z_{\mathrm{in},w}^{(s)} \right|^2}{\sigma_{\mathrm{N}}^2}. \quad (A.2)$$

To evaluate the expression in (A.1), the partial derivatives of (A.2) with respect to the elements of the agent position vector $\mathbf{p}_0$ are calculated as

$$\frac{\partial \ln \Pr\left(O_w^{(s)} \mid \mathbf{p}_0\right)}{\partial x_0} = \frac{1}{\sigma_{\mathrm{N}}^2} \left(\left(O_w^{(s)} - Z_{\mathrm{in},w}^{(s)}\right)^* \cdot \frac{\partial Z_{\mathrm{in},w}^{(s)}}{\partial x_0} + \left(O_w^{(s)} - Z_{\mathrm{in},w}^{(s)}\right) \cdot \frac{\partial \left(Z_{\mathrm{in},w}^{(s)}\right)^*}{\partial x_0} \right), \text{ and}$$

$$(A.3)$$

$$\frac{\partial \ln \Pr\left(O_w^{(s)} \mid \mathbf{p}_0\right)}{\partial y_0} = \frac{1}{\sigma_{\mathrm{N}}^2} \left(\left(O_w^{(s)} - Z_{\mathrm{in},w}^{(s)}\right)^* \cdot \frac{\partial Z_{\mathrm{in},w}^{(s)}}{\partial y_0} + \left(O_w^{(s)} - Z_{\mathrm{in},w}^{(s)}\right) \cdot \frac{\partial \left(Z_{\mathrm{in},w}^{(s)}\right)^*}{\partial y_0} \right). \quad (A.4)$$

Evaluating the product terms in (A.1), we find

$$
\left(\frac{\partial \ln \Pr \left(O_w^{(s)} \mid \mathbf{p}_0 \right)}{\partial x_0} \right) \left(\frac{\partial \ln \Pr \left(O_w^{(s)} \mid \mathbf{p}_0 \right)}{\partial x_0} \right)
$$

$$
= \frac{1}{\sigma_N^4} \left(\frac{\partial Z_{\mathrm{in},w}^{(s)}}{\partial x_0} \cdot \underbrace{\mathsf{E} \left[\left(O_w^{(s)} - Z_{\mathrm{in},w}^{(s)} \right)^* \left(O_w^{(s)} - Z_{\mathrm{in},w}^{(s)} \right)^* \right]}_{=0} \cdot \frac{\partial Z_{\mathrm{in},w}^{(s)}}{\partial x_0} \right.
$$

$$
+ \frac{\partial Z_{\mathrm{in},w}^{(s)}}{\partial x_0} \cdot \underbrace{\mathsf{E} \left[\left(O_w^{(s)} - Z_{\mathrm{in},w}^{(s)} \right)^* \left(O_w^{(s)} - Z_{\mathrm{in},w}^{(s)} \right) \right]}_{=\sigma_N^2} \cdot \frac{\partial \left(Z_{\mathrm{in},w}^{(s)} \right)^*}{\partial x_0}
$$

$$
+ \frac{\partial \left(Z_{\mathrm{in},w}^{(s)} \right)^*}{\partial x_0} \cdot \underbrace{\mathsf{E} \left[\left(O_w^{(s)} - Z_{\mathrm{in},w}^{(s)} \right) \left(O_w^{(s)} - Z_{\mathrm{in},w}^{(s)} \right)^* \right]}_{=\sigma_N^2} \cdot \frac{\partial Z_{\mathrm{in},w}^{(s)}}{\partial x_0}
$$

$$
\left. + \frac{\partial \left(Z_{\mathrm{in},w}^{(s)} \right)^*}{\partial x_0} \cdot \underbrace{\mathsf{E} \left[\left(O_w^{(s)} - Z_{\mathrm{in},w}^{(s)} \right) \left(O_w^{(s)} - Z_{\mathrm{in},w}^{(s)} \right) \right]}_{=0} \cdot \frac{\partial \left(Z_{\mathrm{in},w}^{(s)} \right)^*}{\partial x_0} \right)
$$

$$
= \frac{2}{\sigma_N^2} \left(\frac{\partial Z_{\mathrm{in},w}^{(s)}}{\partial x_0} \cdot \frac{\partial \left(Z_{\mathrm{in},w}^{(s)} \right)^*}{\partial x_0} \right), \tag{A.5}
$$

whereas the remaining terms are calculated equivalently as

$$
\left(\frac{\partial \ln \Pr \left(O_w^{(s)} \mid \mathbf{p}_0 \right)}{\partial y_0} \right) \left(\frac{\partial \ln \Pr \left(O_w^{(s)} \mid \mathbf{p}_0 \right)}{\partial y_0} \right) = \frac{2}{\sigma_N^2} \left(\frac{\partial Z_{\mathrm{in},w}^{(s)}}{\partial y_0} \cdot \frac{\partial \left(Z_{\mathrm{in},w}^{(s)} \right)^*}{\partial y_0} \right), \tag{A.6}
$$

$$
\left(\frac{\partial \ln \Pr \left(O_w^{(s)} \mid \mathbf{p}_0 \right)}{\partial x_0} \right) \left(\frac{\partial \ln \Pr \left(O_w^{(s)} \mid \mathbf{p}_0 \right)}{\partial y_0} \right) = \left(\frac{\partial \ln \Pr \left(O_w^{(s)} \mid \mathbf{p}_0 \right)}{\partial x_0} \right) \left(\frac{\partial \ln \Pr \left(O_w^{(s)} \mid \mathbf{p}_0 \right)}{\partial y_0} \right)
$$

$$
= \frac{1}{\sigma_N^2} \left(\frac{\partial Z_{\mathrm{in},w}^{(s)}}{\partial x_0} \cdot \frac{\partial \left(Z_{\mathrm{in},w}^{(s)} \right)^*}{\partial y_0} + \frac{\partial Z_{\mathrm{in},w}^{(s)}}{\partial y_0} \cdot \frac{\partial \left(Z_{\mathrm{in},w}^{(s)} \right)^*}{\partial x_0} \right). \tag{A.7}
$$

The partial derivatives of the input impedance in (A.5)-(A.7) are found from the impedance expression in (7.1):

$$\frac{\partial Z_{\text{in},w}^{(s)}}{\partial x_0} = \frac{\partial}{\partial x_0} \left(\mathbf{Z}_{\text{M},11} - \mathbf{Z}_{\text{M},12} \cdot \left(\mathbf{Z}_{\text{M},22} + \begin{bmatrix} \mathbf{Z}_{\text{L}}^{(s)} & \mathbf{0} \\ \mathbf{0} & \mathbf{Z}_{\text{C}} \end{bmatrix} \right)^{-1} \cdot \mathbf{Z}_{\text{M},21} \right)$$

$$= -\mathbf{Z}_{\text{M},12} \cdot \frac{\partial}{\partial x_0} \left\{ \left(\mathbf{Z}_{\text{M},22} + \begin{bmatrix} \mathbf{Z}_{\text{L}}^{(s)} & \mathbf{0} \\ \mathbf{0} & \mathbf{Z}_{\text{C}} \end{bmatrix} \right)^{-1} \right\} \cdot \mathbf{Z}_{\text{M},21}. \tag{A.8}$$

All matrices in (A.8) with the exception of $\mathbf{Z}_{\text{L}}^{(s)}$ are dependent on frequency, however, the frequency index w has been omitted for notational convenience. Using the identity $\frac{\partial \mathbf{A}^{-1}}{\partial x} = -\mathbf{A}^{-1}\frac{\partial \mathbf{A}}{\partial x}\mathbf{A}^{-1}$ [114], we find

$$\frac{\partial Z_{\text{in},w}^{(s)}}{\partial x_0} = \mathbf{Z}_{\text{M},12} \cdot \left(\mathbf{Z}_{\text{M},22} + \begin{bmatrix} \mathbf{Z}_{\text{L}}^{(s)} & \mathbf{0} \\ \mathbf{0} & \mathbf{Z}_{\text{C}} \end{bmatrix} \right)^{-1}$$

$$\cdot \begin{bmatrix} \mathbf{0} & \mathbf{0} \\ \mathbf{0} & \frac{\partial \mathbf{Z}_{\text{C}}}{\partial x_0} \end{bmatrix} \cdot \left(\mathbf{Z}_{\text{M},22} + \begin{bmatrix} \mathbf{Z}_{\text{L}}^{(s)} & \mathbf{0} \\ \mathbf{0} & \mathbf{Z}_{\text{C}} \end{bmatrix} \right)^{-1} \cdot \mathbf{Z}_{\text{M},21}, \tag{A.9}$$

where the only nonzero elements of the partial derivative of the coupling impedance matrix $\mathbf{Z}_{\text{C}}$—as defined in (3.47)—are those that vary as a function of the agent position. These elements are found in only one row and column of $\mathbf{Z}_{\text{C}}$, namely the ones pertaining to the pairwise coupling coefficients of the agent with all anchors. By denoting the index of this row and column as $\tilde{n}$, we can formally state the derivatives as

$$\left[\frac{\partial \mathbf{Z}_{\text{C}}}{\partial x_0} \right]_{n\tilde{n}} = \left[\frac{\partial \mathbf{Z}_{\text{C}}}{\partial x_0} \right]_{\tilde{n}n} = \begin{cases} j\omega_w \frac{\partial M_n}{\partial x_0} & : \ n \neq \tilde{n}, \\ 0 & : \ n = \tilde{n}, \end{cases} \tag{A.10}$$

with M_n denoting the mutual inductance between the agent and the nth anchor, which may be either the measuring or a passive anchor. By expanding the mutual inductance terms using the dipole approximation (cf. (3.20)) and noting that the polarization factor J_n is equal to 1 in the assumed coplanar case, we obtain

$$\left[\frac{\partial \mathbf{Z}_{\text{C}}}{\partial x_0} \right]_{\tilde{n}n} = -j\omega_w \cdot \frac{3}{4} \frac{\mu \pi \nu_n \nu_0 r_n^2 r_0^2 \left(x_n - x_0 \right)}{\sqrt{\left(x_n - x_0 \right)^2 + \left(y_n - y_0 \right)^2}^5} \qquad \forall n \neq \tilde{n}. \tag{A.11}$$

The derivative of the conjugate input impedance, on the other hand, follows from taking the complex conjugate of (A.8):

$$
\frac{\partial \left(Z_{\text{in},w}^{(s)} \right)^*}{\partial x_0} = \mathbf{Z}_{\text{M},12}^* \cdot \left(\mathbf{Z}_{\text{M},22}^* + \begin{bmatrix} Z_{\text{L}}^{(s)} & 0 \\ 0 & Z_{\text{C}} \end{bmatrix}^* \right)^{-1} \cdot
$$
$$
\begin{bmatrix} 0 & 0 \\ 0 & \frac{\partial Z_{\text{C}}^*}{\partial x_0} \end{bmatrix} \cdot \left(\mathbf{Z}_{\text{M},22}^* + \begin{bmatrix} Z_{\text{L}} & 0 \\ 0 & Z_{\text{C}} \end{bmatrix}^* \right)^{-1} \cdot \mathbf{Z}_{\text{M},21}^*, \tag{A.12}
$$

With $\mathfrak{Re}\{\mathbf{Z}_{\text{C}}\} = \mathbf{0}$, we find

$$
\frac{\partial \left(\mathbf{Z}_{\text{C}} \right)^*}{\partial x_0} = -\frac{\partial \mathbf{Z}_{\text{C}}}{\partial x_0}. \tag{A.13}
$$

The calculations above follow equivalently for the partial derivatives with respect to y_0.

Acronyms

AF   Amplify-and-forward.

ANN   Artificial neural network.

AOA   Angle of arrival.

AWGN Additive white Gaussian noise.

CDF   Cumulative distribution function.

CF   Compress-and-forward.

CNT   Carbon nanotube.

CRLB Cramér-Rao lower bound.

CSI   Channel state information.

DC   Direct current.

DF   Decode-and-forward.

DOF   Degrees of freedom.

DUT   Device under test.

FDTD Finite difference time domain method.

FEM   Finite element method.

i.i.d. Independent and identically distributed.

IF   Intermediate frequency.

LNA   Low-noise amplifier.

MEMS Microelectromechanical systems.

MOM	Method of moments.
MQS	Magnetoquasistatic.
NEMS	Nanoelectromechanical systems.
NFC	Near-field communication.
NRA	Noise of the receiver antenna.
NRC	Noise of the receiver circuit.
NTA	Noise of the transmitter antenna.
NTC	Noise of the transmitter circuit.
PCB	Printed circuit board.
PDF	Probability distribution function.
PSD	Power spectral density.
RF	Radio frequency.
RFID	Radio frequency identification.
RMS	Root mean square.
RMSE	Root mean squared error.
RSS	Received signal strength.
SINR	Signal-to-interference-plus-noise ratio.
SNR	Signal-to-noise ratio.
SQUID	Superconducting Quantum Interference Device.
TOA	Time of arrival.
UWB	Ultra-wideband.
VNA	Vector network analyzer.
XOR	Exclusive OR.

Notation

General Definitions

a, A	Scalars a and A.
$\mathbf{a}$	Vector $\mathbf{a}$.
$\mathbf{A}$	Matrix $\mathbf{A}$.
$\mathbf{A}[n,:]$	Row n of Matrix $\mathbf{A}$.
$\mathbf{A}[:,n]$	Column n of Matrix $\mathbf{A}$.
$[\mathbf{a}]_n$	Element n of vector $\mathbf{a}$.
$[\mathbf{A}]_{mn}$	Element (m,n) of matrix $\mathbf{A}$.
$\breve{a}$	Trial value of a.
$\hat{a}$	Estimate value of a.

Operators, Measures, and Functions

$\lvert\mathcal{A}\rvert$	Cardinality of set $\mathcal{A}$.
$(\cdot)^*$	Complex conjugate.
$\lvert\mathbf{A}\rvert$	Determinant of matrix $\mathbf{A}$.
$\operatorname{diag}\{\cdot\}$	Diagonal matrix from elements of argument.
$\lVert\cdot\rVert$	Euclidean norm of vector.
$\exp(\cdot)$	Natural exponential function.
$\mathsf{E}[\cdot]$	Expectation of random variable.
$\lVert\cdot\rVert_{\mathrm{F}}$	Frobenius norm of matrix.

$\odot$	Hadamard product.
$(\cdot)^{\mathrm{H}}$	Hermitian transpose of vector or matrix.
$\mathfrak{Im}\left\{\cdot\right\}$	Imaginary part.
$(\cdot)^{-1}$	Matrix inverse.
δ_{mn}	Kronecker delta function.
$\otimes$	Kronecker product.
∇	Nabla operator.
$\mathbf{A} \succeq \mathbf{B}$	$\mathbf{A} - \mathbf{B}$ is positive semidefinite.
$\mathrm{Pr}\left(\cdot\right)$	Probability.
$(\cdot)^{+}$	Moore-Penrose pseudoinverse.
$\mathfrak{Re}\left\{\cdot\right\}$	Real part.
$\mathrm{sgn}\left\{\cdot\right\}$	Sign function.
$\mathrm{tr}\left\{\cdot\right\}$	Trace of matrix.
$(\cdot)^{\mathrm{T}}$	Transpose of vector or matrix.

Variables

$\mathbf{1}_n$	All-ones vector of dimension n.
a	Antenna wire radius.
a_n	Power wave entering port n.
$\mathbf{a}$	Surface area vector.
$\mathbf{A}$	Magnetic vector potential.
b	Side length of bounding box for localization.
b_n	Power wave leaving port n.
$\mathbf{B}$	Magnetic flux density (vector).
C	Capacitance.
$\mathbf{c}$	Vector of current message bits for joint decoding and localization.
$\mathbf{C}$	Matrix of all message bits for joint decoding and localization.

d	Distance.
δ	Skin depth.
$\mathbf{E}$	Electric field strength (vector).
$\mathbf{E}_n$	Single-entry matrix: 1 at position (n,n), 0 otherwise.
ε	Electric permittivity.
η_0	Free space wave impedance.
f	Frequency.
$f(y,\xi)$	Activation function of neuron with input y and threshold ξ.
Δf	Bandwidth.
$\mathbf{F}$	Fisher information matrix.
G	Voltage gain of linear two-port network.
$\mathbf{g}$	Vector of relay gains.
Γ	Reflection coefficient.
h	Channel coefficient.
h_{P}	Planck's constant.
$\mathbf{H}$	Channel matrix.
i	Current.
$\mathbf{I}$	Identity matrix.
J	Polarization factor.
$\mathbf{J}$	Current density (vector).
k	Coupling coefficient.
k_{B}	Boltzmann constant, $k_{\mathrm{B}} = 1.38064852 \cdot 10^{-23}\,\mathrm{J} \cdot K^{-1}$.
$\mathbf{k}$	Vector of coupling coefficients.
L	Self inductance.
λ	Wavelength.
Λ	Likelihood ratio.
M	Mutual inductance.
$\mathbf{m}_{\mathrm{d}}$	Magnetic dipole moment.

μ	Magnetic permeability.
N_0	Noise power spectral density.
N_A	Number of anchors.
N_{ex}	Number of excess relays.
N_{freq}	Number of frequency samples.
N_S	Number of secondary nodes.
ν	Number of antenna windings.
O	Noisy measurement observation.
ω	Angular frequency, $\omega = 2\pi f$.
ω_{res}	Resonance frequency.
Ω	Volume.
P	Number of measurement points for drift autocalibration.
P_T	Transmit power.
$\mathbf{p}$	Node position vector.
$\mathbf{p_0}$	Position vector of agent node.
Φ	Magnetic flux contribution of a single loop winding.
φ	Azimuth (spherical coordinates).
Ψ	Magnetic flux.
Q	Quality factor.
$\mathbf{q}$	Node orientation vector.
$\mathbf{Q}$	Weight matrix of artificial neural network.
r	Antenna radius.
R	Resistance.
R_Ω	Loss resistance.
R_R	Radiation resistance.
$\mathbf{r}$	Point in space.
ρ	Radius (spherical coordinates).
ρ_Q	Charge density.

ρ_C	Correlation coefficient.
S	Number of load switching stages.
$\mathbf{S}$	Scattering parameter matrix.
σ	Electrical conductivity.
σ_N^2	Noise variance.
Σ	Surface.
$\delta\Sigma$	Boundary of surface Σ.
$\boldsymbol{\Sigma}$	Covariance matrix.
T	Temperature.
t	Time.
$\boldsymbol{\theta}$	Parameter vector containing node position and orientation.
$\boldsymbol{\theta}_0$	Parameter vector of agent node.
ϑ	Inclination (spherical coordinates).
u	Voltage.
x	Transmitted signal.
$\Xi(\mathbf{p}_0)$	Reliability metric for joint decoding and localization.
y	Received signal.
Y	Admittance.
Z	Impedance.
Z_0	Characteristic impedance.
Z_{in}	Input impedance.
$\mathbf{0}$	Zero matrix.

Bibliography

[1] J. Agre and L. Clare, "An integrated architecture for cooperative sensing networks," *Computer*, vol. 33, no. 5, pp. 106–108, 2000.

[2] D. Ahn and S. Hong, "A study on magnetic field repeater in wireless power transfer," *Industrial Electronics, IEEE Transactions on*, vol. 60, no. 1, pp. 360–371, 2013.

[3] I. F. Akyildiz, F. Brunetti, and C. Blázquez, "Nanonetworks: A new communication paradigm," *Computer Networks*, vol. 52, no. 12, pp. 2260–2279, 2008.

[4] I. F. Akyildiz and J. M. Jornet, "Electromagnetic wireless nanosensor networks," *Nano Communication Networks*, vol. 1, no. 1, pp. 3–19, 2010.

[5] I. F. Akyildiz, W. Su, Y. Sankarasubramaniam, and E. Cayirci, "A survey on sensor networks," *Communications magazine, IEEE*, vol. 40, no. 8, pp. 102–114, 2002.

[6] I. F. Akyildiz, Z. Sun, and M. C. Vuran, "Signal propagation techniques for wireless underground communication networks," *Physical Communication*, vol. 2, no. 3, pp. 167–183, 2009.

[7] P. E. Allen and E. Sanchez-Sinencio, *Switched capacitor circuits.* Van Nostrand Reinhold, 1984.

[8] L. Angrisani, F. Bonavolontà, G. d'Alessandro, and M. D'Arco, "Inductive power transmission for wireless sensor networks supply," in *Environmental Energy and Structural Monitoring Systems (EESMS), 2014 IEEE Workshop on.* IEEE, 2014, pp. 1–5.

[9] J. N. Ash and R. L. Moses, "On optimal anchor node placement in sensor localization by optimization of subspace principal angles," in *Acoustics, Speech and Signal Processing, 2008. ICASSP 2008. IEEE International Conference on.* IEEE, 2008, pp. 2289–2292.

[10] S. Babic and C. Akyel, "New analytic-numerical solutions for the mutual inductance of two coaxial circular coils with rectangular cross section in air," *Magnetics, IEEE Transactions on*, vol. 42, no. 6, pp. 1661–1669, 2006.

[11] S. Babic, F. Sirois, C. Akyel, and C. Girardi, "Mutual inductance calculation between circular filaments arbitrarily positioned in space: alternative to grover's formula," *Magnetics, IEEE Transactions on*, vol. 46, no. 9, pp. 3591–3600, 2010.

[12] C. A. Balanis, *Antenna theory: analysis and design*. John Wiley & Sons, 2005.

[13] S. Basu, Y. Gerchman, C. H. Collins, F. H. Arnold, and R. Weiss, "A synthetic multicellular system for programmed pattern formation," *Nature*, vol. 434, no. 7037, pp. 1130–1134, 2005.

[14] R. W. Beatty, "Some basic microwave phase shift equations," *Radio Sci. J. Res., NBS/USNC-URSI*, vol. 6, pp. 349–353, 1964.

[15] W. R. Bennett, "A general review of linear varying parameter and nonlinear circuit analysis," *Proceedings of the IRE*, vol. 38, no. 3, pp. 259–263, March 1950.

[16] H. Bölcskei, R. U. Nabar, Ö. Oyman, and A. J. Paulraj, "Capacity scaling laws in MIMO relay networks," *Wireless Communications, IEEE Transactions on*, vol. 5, no. 6, pp. 1433–1444, 2006.

[17] B. Bougard, F. Catthoor, D. C. Daly, A. Chandrakasan, and W. Dehaene, "Energy efficiency of the IEEE 802.15. 4 standard in dense wireless microsensor networks: Modeling and improvement perspectives," in *Proceedings of the conference on Design, Automation and Test in Europe-Volume 1*. IEEE Computer Society, 2005, pp. 196–201.

[18] B. Braem, B. Latre, I. Moerman, C. Blondia, E. Reusens, W. Joseph, L. Martens, and P. Demeester, "The need for cooperation and relaying in short-range high path loss sensor networks," in *Sensor Technologies and Applications, 2007. SensorComm 2007. International Conference on*. IEEE, 2007, pp. 566–571.

[19] H.-D. Brüns, C. Schuster, and H. Singer, "Numerical electromagnetic field analysis for EMC problems," *Electromagnetic Compatibility, IEEE Transactions on*, vol. 49, no. 2, pp. 253–262, 2007.

[20] G. H. Bryant, *Principles of microwave measurements*. IET, 1993, vol. 5.

[21] B. L. Cannon, J. F. Hoburg, D. D. Stancil, and S. C. Goldstein, "Magnetic resonant coupling as a potential means for wireless power transfer to multiple small receivers," *Power Electronics, IEEE Transactions on*, vol. 24, no. 7, pp. 1819–1825, 2009.

[22] C. W. Chan and C. J. Stevens, "Two-dimensional magneto-inductive wave data structures," in *Antennas and Propagation (EUCAP), Proceedings of the 5th European Conference on*. IEEE, 2011, pp. 1071–1075.

[23] A. Chandrakasan, R. Amirtharajah, S. Cho, J. Goodman, G. Konduri, J. Kulik, W. Rabiner, and A. Wang, "Design considerations for distributed microsensor systems," in *Custom Integrated Circuits, 1999. Proceedings of the IEEE 1999*. IEEE, 1999, pp. 279–286.

[24] J. Choma, *Electrical networks: theory and analysis.* Wiley-Interscience, 1985.

[25] G. Ciuti, P. Valdastri, A. Menciassi, and P. Dario, "Robotic magnetic steering and locomotion of capsule endoscope for diagnostic and surgical endoluminal procedures," *Robotica*, vol. 28, no. 02, pp. 199–207, 2010.

[26] V. Coskun, B. Ozdenizci, and K. Ok, "A survey on near field communication (NFC) technology," *Wireless personal communications*, vol. 71, no. 3, pp. 2259–2294, 2013.

[27] G. Cybenko, "Approximation by superpositions of a sigmoidal function," *Mathematics of control, signals and systems*, vol. 2, no. 4, pp. 303–314, 1989.

[28] H. Davis, "The analysis and synthesis of a class of linear time-varying networks," *Transactions of the American Institute of Electrical Engineers, Part II: Applications and Industry*, vol. 81, no. 6, pp. 325–330, Jan 1963.

[29] R. Dengler, "Self inductance of a wire loop as a curve integral," *arXiv preprint arXiv:1204.1486*, 2012.

[30] G. Dogangil, O. Ergeneman, J. J. Abbott, S. Pané, H. Hall, S. Muntwyler, and B. J. Nelson, "Toward targeted retinal drug delivery with wireless magnetic microrobots," in *Intelligent Robots and Systems, 2008. IROS 2008. IEEE/RSJ International Conference on.* IEEE, 2008, pp. 1921–1926.

[31] M. C. Domingo, "Magnetic induction for underwater wireless communication networks," *Antennas and Propagation, IEEE Transactions on*, vol. 60, no. 6, pp. 2929–2939, 2012.

[32] D. Drung, C. Assmann, J. Beyer, A. Kirste, M. Peters, F. Ruede, and T. Schurig, "Highly sensitive and easy-to-use SQUID sensors," *Applied Superconductivity, IEEE Transactions on*, vol. 17, no. 2, pp. 699–704, 2007.

[33] G. Dumphart, "System model for magnetoinductive communication," Communication Technology Laboratory, ETH Zurich, Tech. Rep., 2015.

[34] G. Dumphart and A. Wittneben, "Stochastic misalignment model for magneto-inductive SISO and MIMO links," in *IEEE International Symposium on Personal, Indoor and Mobile Radio Communications (PIMRC)*, Sep. 2016.

[35] A. W. Eckford, "Achievable information rates for molecular communication with distinct molecules," in *Bio-Inspired Models of Network, Information and Computing Systems, 2007. Bionetics 2007. 2nd.* IEEE, 2007, pp. 313–315.

[36] ——, "Nanoscale communication with Brownian motion," in *Information Sciences and Systems, 2007. CISS'07. 41st Annual Conference on.* IEEE, 2007, pp. 160–165.

[37] J. Engberg and T. Larsen, *Noise theory of linear and nonlinear circuits*. John Wiley & Sons, 1995.

[38] O. Ergeneman, G. Chatzipirpiridis, J. Pokki, M. Marín-Suárez, G. A. Sotiriou, S. Medina-Rodríguez, J. F. F. Sánchez, A. Fernández-Gutiérrez, S. Pane, and B. J. Nelson, "In vitro oxygen sensing using intraocular microrobots," *Biomedical Engineering, IEEE Transactions on*, vol. 59, no. 11, pp. 3104–3109, 2012.

[39] M. M. Eshaghian-Wilner, A. Friesz, A. Khitun, S. Navab, A. C. Parker, K. L. Wang, and C. Zhou, "Emulation of neural networks on a nanoscale architecture," in *Journal of Physics: Conference Series*, vol. 61, no. 1. IOP Publishing, 2007, p. 288.

[40] H.-J. Eul and B. Schiek, "A generalized theory and new calibration procedures for network analyzer self-calibration," *Microwave Theory and Techniques, IEEE Trans. on*, vol. 39, no. 4, pp. 724–731, 1991.

[41] C. Falconi, A. D'Amico, and Z. L. Wang, "Wireless joule nanoheaters," *Sensors and Actuators B: Chemical*, vol. 127, no. 1, pp. 54–62, 2007.

[42] K. Finkenzeller, *RFID handbook*, 3rd ed. John Wiley & Sons, Ltd, 2010.

[43] R. H. Fowler and L. Nordheim, "Electron emission in intense electric fields," in *Proceedings of the Royal Society of London A: Mathematical, Physical and Engineering Sciences*, vol. 119, no. 781. The Royal Society, 1928, pp. 173–181.

[44] H. T. Friis, "A note on a simple transmission formula," *Proceedings of the IRE*, vol. 34, no. 5, pp. 254–256, 1946.

[45] P. K. Gentner, R. Langwieser, A. L. Scholtz, G. Hofer, and C. F. Mecklenbräuker, "A UHF/UWB hybrid silicon RFID tag with on–chip antennas," *EURASIP Journal on Embedded Systems*, vol. 2013, no. 1, pp. 1–12, 2013.

[46] S. C. Goldstein, J. D. Campbell, and T. C. Mowry, "Programmable matter," *Computer*, vol. 38, no. 6, pp. 99–101, 2005.

[47] L. Greengard, *The rapid evaluation of potential fields in particle systems*. MIT press, 1988.

[48] F. W. Grover, "The calculation of the mutual inductance of circular filaments in any desired positions," *Proceedings of the IRE*, vol. 32, no. 10, pp. 620–629, 1944.

[49] ——, *Inductance calculations: working formulas and tables*. Dover, New York, 1962.

[50] I. Hammerström, M. Kuhn, and A. Wittneben, "Impact of relay gain allocation on the performance of cooperative diversity networks," in *Vehicular Technology Conference, 2004. VTC2004-Fall. 2004 IEEE 60th*, vol. 3. IEEE, 2004, pp. 1815–1819.

[51] G. W. Hanson, "Fundamental transmitting properties of carbon nanotube antennas," *Antennas and Propagation, IEEE Transactions on*, vol. 53, no. 11, pp. 3426–3435, 2005.

[52] S. Hashi, Y. Tokunaga, S. Yabukami, T. Kohno, T. Ozawa, Y. Okazaki, K. Ishiyama, and K. Arai, "Wireless motion capture system using magnetically coupled LC resonant marker," *Journal of magnetism and magnetic materials*, vol. 290, pp. 1330–1333, 2005.

[53] H. A. Haus and J. R. Melcher, *Electromagnetic fields and energy*. Prentice Hall, 1989.

[54] S. Haykin, *Neural networks: a comprehensive foundation*, 2nd ed. Prentice Hall, Jul. 1998.

[55] W. H. Hayt and J. A. Buck, *Engineering electromagnetics*, 8th ed. McGraw-Hill New York, 2012.

[56] M. Hiebel, "Vector network analyzer (VNA) calibration: the basics," Rohde & Schwarz, White Paper, 2008.

[57] A. Hirose, *Complex-valued neural networks: theories and applications*. World Scientific, 2003, vol. 5.

[58] S. Hiyama, Y. Moritani, T. Suda, R. Egashira, A. Enomoto, and T. Nakano, "Molecular Communication," in *Proc. of the 2005 NSTI Nanotechnology Conference*, vol. 3, May 2005, pp. 392–395.

[59] E. Hökenek and G. Moschytz, "Analysis of multiphase switched-capacitor (m.s.c.) networks using the indefinite admittance matrix (i.q.m.)," in *IEE Proceedings G-Electronic Circuits and Systems*, vol. 127, no. 5. IET, 1980, pp. 226–241.

[60] F. Hooge, "$1/f$ noise sources," *Electron Devices, IEEE Transactions on*, vol. 41, no. 11, pp. 1926–1935, 1994.

[61] K. Hornik, M. Stinchcombe, and H. White, "Multilayer feedforward networks are universal approximators," *Neural networks*, vol. 2, no. 5, pp. 359–366, 1989.

[62] M. T. Hou, H.-M. Shen, G.-L. Jiang, C.-N. Lu, I.-J. Hsu, J. A. Yeh *et al.*, "A rolling locomotion method for untethered magnetic microrobots," *Applied Physics Letters*, vol. 96, no. 2, p. 024102, 2010.

[63] T. H. Hubing, "Survey of numerical electromagnetic modeling techniques," Department of Electrical Engineering, University of Missouri-Rolla, USA, Tech. Rep., 1991.

[64] N. S. Hudak and G. G. Amatucci, "Small-scale energy harvesting through thermoelectric, vibration, and radiofrequency power conversion," *Journal of Applied Physics*, vol. 103, no. 10, p. 101301, 2008.

[65] G. Iddan, G. Meron, A. Glukhovsky, and P. Swain, "Wireless capsule endoscopy," *Nature*, vol. 405, p. 417, 2000.

[66] *Near field communication - interface and protocol (NFCIP-1)*, ISO/IEC Std. 18 092:2013, 2013.

[67] *Near field communication - interface and protocol (NFCIP-2)*, ISO/IEC Std. 21 481:2012, 2013.

[68] M. T. Ivrlač, J. Nossek *et al.*, "Toward a circuit theory of communication," *Circuits and Systems I: Regular Papers, IEEE Transactions on*, vol. 57, no. 7, pp. 1663–1683, 2010.

[69] K. Jensen, J. Weldon, H. Garcia, and A. Zettl, "Nanotube radio," *Nano letters*, vol. 7, no. 11, pp. 3508–3511, 2007.

[70] B. Jiang, J. R. Smith, M. Philipose, S. Roy, K. Sundara-Rajan, and A. V. Mamishev, "Energy scavenging for inductively coupled passive RFID systems," *Instrumentation and Measurement, IEEE Transactions on*, vol. 56, no. 1, pp. 118–125, 2007.

[71] J. M. Kahn, R. H. Katz, and K. S. Pister, "Next century challenges: mobile networking for 'smart dust'," in *Proceedings of the 5th annual ACM/IEEE international conference on Mobile computing and networking*. ACM, 1999, pp. 271–278.

[72] M. Kamon, M. J. Tsuk, and J. K. White, "FASTHENRY: A multipole-accelerated 3-D inductance extraction program," *Microwave Theory and Techniques, IEEE Transactions on*, vol. 42, no. 9, pp. 1750–1758, 1994.

[73] A. Karalis, J. D. Joannopoulos, and M. Soljačić, "Efficient wireless non-radiative mid-range energy transfer," *Annals of Physics*, vol. 323, no. 1, pp. 34–48, 2008.

[74] L. B. Kish, "End of Moore's law: thermal (noise) death of integration in micro and nano electronics," *Physics Letters A*, vol. 305, no. 3, pp. 144–149, 2002.

[75] S. Kisseleff, I. Akyildiz, and W. Gerstacker, "Interference polarization in magnetic induction based Wireless Underground Sensor Networks," in *Personal, Indoor and Mobile Radio Communications (PIMRC Workshops), 2013 IEEE 24th International Symposium on*, Sept 2013, pp. 71–75.

[76] ——, "Digital signal transmission in magnetic induction based wireless underground sensor networks," *Communications, IEEE Transactions on*, vol. 63, no. 6, pp. 2300–2311, June 2015.

[77] S. Kisseleff, W. Gerstacker, R. Schober, Z. Sun, and I. F. Akyildiz, "Channel capacity of magnetic induction based wireless underground sensor networks under practical

constraints," in *Wireless Communications and Networking Conference (WCNC), 2013 IEEE*. IEEE, 2013, pp. 2603–2608.

[78] A. Kulakov, D. Davcev, and G. Trajkovski, "Application of wavelet neural-networks in wireless sensor networks," in *Software Engineering, Artificial Intelligence, Networking and Parallel/Distributed Computing, 2005 and First ACIS International Workshop on Self-Assembling Wireless Networks. SNPD/SAWN 2005. Sixth International Conference on*. IEEE, 2005, pp. 262–267.

[79] K. Kurokawa, "Power waves and the scattering matrix," *Microwave Theory and Techniques, IEEE Transactions on*, vol. 13, no. 2, pp. 194–202, 1965.

[80] A. Kurs, A. Karalis, R. Moffatt, J. D. Joannopoulos, P. Fisher, and M. Soljačić, "Wireless power transfer via strongly coupled magnetic resonances," *science*, vol. 317, no. 5834, pp. 83–86, 2007.

[81] C. F. Kurth and G. S. Moschytz, "Nodal analysis of switched-capacitor networks," *Circuits and Systems, IEEE Transactions on*, vol. 26, no. 2, pp. 93–105, 1979.

[82] C. K. Lee, W. Zhong, and S. Hui, "Effects of magnetic coupling of nonadjacent resonators on wireless power domino-resonator systems," *Power Electronics, IEEE Transactions on*, vol. 27, no. 4, pp. 1905–1916, 2012.

[83] B. Lenaerts and R. Puers, "Inductive powering of a freely moving system," *Sensors and Actuators A: Physical*, vol. 123, pp. 522–530, 2005.

[84] H. Li, G. Yan, P. Jiang, and P. Zan, "A portable electromagnetic localization method for micro devices in vivo," in *Automation Congress, 2008. WAC 2008. World*. IEEE, 2008, pp. 1–4.

[85] K. Li, H. Luan, and C.-C. Shen, "Qi-ferry: Energy-constrained wireless charging in wireless sensor networks," in *2012 IEEE Wireless Communications and Networking Conference (WCNC)*. IEEE, 2012, pp. 2515–2520.

[86] M. Li, Y. Liu, and L. Chen, "Nonthreshold-based event detection for 3D environment monitoring in sensor networks," *Knowledge and Data Engineering, IEEE Transactions on*, vol. 20, no. 12, pp. 1699–1711, 2008.

[87] K. R. Liu, *Cooperative communications and networking*. Cambridge university press, 2009.

[88] K. J. Loh, J. P. Lynch, and N. A. Kotov, "Inductively coupled nanocomposite wireless strain and pH sensors," *Smart Structures and Systems*, vol. 4, no. 5, pp. 531–548, 2008.

[89] H. M. Lu, C. Goldsmith, L. Cauller, and J.-B. Lee, "MEMS-based inductively coupled RFID transponder for implantable wireless sensor applications," *Magnetics, IEEE Transactions on*, vol. 43, no. 6, pp. 2412–2414, 2007.

[90] H.-C. Lu and T.-H. Chu, "Port reduction methods for scattering matrix measurement of an *n*-port network," *IEEE Transactions on Microwave Theory and Techniques*, vol. 48, no. 6, pp. 959–968, 2000.

[91] S. Lupi, M. Forzan, and A. Aliferov, *Induction and direct resistance heating.* Springer, 2015.

[92] S. A. Maas, *Noise in linear and nonlinear circuits.* Artech House Publishers, 2005.

[93] G. Mao, B. Fidan, and B. D. Anderson, "Wireless sensor network localization techniques," *Computer networks*, vol. 51, no. 10, pp. 2529–2553, 2007.

[94] A. Markham and N. Trigoni, "Magneto-inductive networked rescue system (MINERS): taking sensor networks underground," in *Proceedings of the 11th international conference on Information Processing in Sensor Networks.* ACM, 2012, pp. 317–328.

[95] A. Markham, N. Trigoni, S. A. Ellwood, and D. W. Macdonald, "Revealing the hidden lives of underground animals using magneto-inductive tracking," in *Proceedings of the 8th ACM Conference on Embedded Networked Sensor Systems.* ACM, 2010, pp. 281–294.

[96] M. Masihpour, "Cooperative communication in near field magnetic induction communication systems," Ph.D. dissertation, University of Technology, Sydney, 2012.

[97] M. Mauve, J. Widmer, and H. Hartenstein, "A survey on position-based routing in mobile ad hoc networks," *Network, IEEE*, vol. 15, no. 6, pp. 30–39, 2001.

[98] W. S. McCulloch and W. Pitts, "A logical calculus of the ideas immanent in nervous activity," *The bulletin of mathematical biophysics*, vol. 5, no. 4, pp. 115–133, 1943.

[99] Z. W. Mekonnen, "Time of arrival based infrastructureless human posture capturing system," Ph.D. dissertation, ETH Zurich, 2016.

[100] Z. W. Mekonnen, E. Slottke, H. Luecken, C. Steiner, and A. Wittneben, "Constrained maximum likelihood positioning for uwb based human motion tracking," in *Indoor Positioning and Indoor Navigation (IPIN), 2010 International Conference on.* IEEE, 2010, pp. 1–10.

[101] G. Moore, "Cramming more components onto integrated circuits," *Electronics*, vol. 38, no. 8, pp. 114–117, 1965.

[102] M. Moore, A. Enomoto, T. Nakano, R. Egashira, T. Suda, A. Kayasuga, H. Kojima, H. Sakakibara, and K. Oiwa, "A design of a molecular communication system for nanomachines using molecular motors," in *Fourth Annual IEEE International Conference on Pervasive Computing and Communications Workshops (PERCOMW'06)*. IEEE, March 2006.

[103] M. J. Moore, T. Suda, and K. Oiwa, "Molecular communication: modeling noise effects on information rate," *NanoBioscience, IEEE Transactions on*, vol. 8, no. 2, pp. 169–180, 2009.

[104] T. Nakano, M. J. Moore, F. Wei, A. V. Vasilakos, and J. Shuai, "Molecular communication and networking: Opportunities and challenges," *NanoBioscience, IEEE Transactions on*, vol. 11, no. 2, pp. 135–148, 2012.

[105] J. A. Nelder and R. Mead, "A simplex method for function minimization," *The computer journal*, vol. 7, no. 4, pp. 308–313, 1965.

[106] B. J. Nelson, I. K. Kaliakatsos, and J. J. Abbott, "Microrobots for minimally invasive medicine," *Annual review of biomedical engineering*, vol. 12, pp. 55–85, 2010.

[107] F. E. Neumann, "Allgemeine Gesetze der inducirten elektrischen Ströme," *Annalen der Physik*, vol. 143, no. 1, pp. 31–44, 1846.

[108] H. Nyquist, "Thermal Agitation of Electric Charge in Conductors," *Phys. Rev.*, vol. 32, pp. 110–113, Jul 1928.

[109] F. Oldewurtel and P. Mähönen, "Neural wireless sensor networks," in *Systems and Networks Communications, 2006. ICSNC'06. International Conference on*. IEEE, 2006, pp. 28–28.

[110] D. Oyama, Y. Adachi, M. Higuchi, and G. Uehara, "Magnetic marker localization system using a super-low-frequency signal," *Magnetics, IEEE Transactions on*, vol. 50, no. 11, pp. 1–4, 2014.

[111] L. Palzer, "Analysis of wireles artificial neural networks based on coherent relaying," 2013, eTH Zurich, semester project report.

[112] E. Paperno, I. Sasada, and E. Leonovich, "A new method for magnetic position and orientation tracking," *Magnetics, IEEE Transactions on*, vol. 37, no. 4, pp. 1938–1940, 2001.

[113] A. Papoulis and S. U. Pillai, *Probability, random variables, and stochastic processes*. McGraw-Hill, 2002.

[114] K. B. Petersen and M. S. Pedersen, "The matrix cookbook," Nov 2012, version 20121115.

[115] G. J. Pottie and W. J. Kaiser, "Wireless integrated network sensors," *Communications of the ACM*, vol. 43, no. 5, pp. 51–58, 2000.

[116] R. A. Potyrailo, W. G. Morris, T. Sivavec, H. W. Tomlinson, S. Klensmeden, and K. Lindh, "RFID sensors based on ubiquitous passive 13.56-MHz RFID tags and complex impedance detection," *Wireless Communications and Mobile Computing*, vol. 9, no. 10, pp. 1318–1330, 2009.

[117] D. M. Pozar, *Microwave engineering.* John Wiley & Sons, 2009.

[118] F. H. Raab, E. B. Blood, T. O. Steiner, and H. R. Jones, "Magnetic position and orientation tracking system," *Aerospace and Electronic Systems, IEEE Transactions on*, vol. AES-15, no. 5, pp. 709–718, 1979.

[119] S. Ramo, J. R. Whinnery, and T. Van Duzer, *Fields and waves in communication electronics.* John Wiley & Sons, 2008.

[120] C. Rohrig and S. Spieker, "Tracking of transport vehicles for warehouse management using a wireless sensor network," in *Intelligent Robots and Systems, 2008. IROS 2008. IEEE/RSJ International Conference on.* IEEE, 2008, pp. 3260–3265.

[121] R. Rojas, *Neural networks: a systematic introduction.* Springer Science & Business Media, 2013.

[122] I. Rolfes and B. Schiek, "Measurement of the scattering-parameters of planar multi-port devices," in *2005 European Microwave Conference*, vol. 2. IEEE, 2005, pp. 4–pp.

[123] R. Rolny, J. Wagner, C. Eşli, and A. Wittneben, "Distributed gain matrix optimization in non-regenerative MIMO relay networks," in *Signals, Systems and Computers, 2009 Conference Record of the Forty-Third Asilomar Conference on.* IEEE, 2009, pp. 1503–1507.

[124] R. T. Rolny, M. Kuhn, and A. Wittneben, "The relay carpet: Ubiquitous two-way relaying in cooperative cellular networks," in *Personal Indoor and Mobile Radio Communications (PIMRC), 2013 IEEE 24th International Symposium on.* IEEE, 2013, pp. 1174–1179.

[125] E. J. Rothwell and M. J. Cloud, *Electromagnetics.* CRC press, 2008.

[126] S. Roundy, D. Steingart, L. Frechette, P. Wright, and J. Rabaey, "Power sources for wireless sensor networks," in *Wireless sensor networks.* Springer, 2004, pp. 1–17.

[127] D. E. Rumelhart, G. E. Hinton, and R. J. Williams, "Learning representations by back-propagating errors," *Cognitive modeling*, vol. 5, no. 3, p. 1, 1988.

[128] D. Rytting, "Network analyzer error models and calibration methods," Agilent Technologies, Tech. Rep., 1998.

[129] D. K. Rytting, "Network analyzer accuracy overview," in *ARFTG Conference Digest-Fall, 58th*, vol. 40. IEEE, 2001, pp. 1–13.

[130] Y. Sasaki, Y. Shioyama, W.-J. Tian, J.-i. Kikuchi, S. Hiyama, Y. Moritani, and T. Suda, "A nanosensory device fabricated on a liposome for detection of chemical signals," *Biotechnology and bioengineering*, vol. 105, no. 1, pp. 37–43, 2010.

[131] A. Savvides, C.-C. Han, and M. B. Strivastava, "Dynamic fine-grained localization in ad-hoc networks of sensors," in *Proceedings of the 7th annual international conference on Mobile computing and networking.* ACM, 2001, pp. 166–179.

[132] P. Scholz, "Analysis and numerical modeling of inductively coupled antenna systems," Ph.D. dissertation, Theorie Elektromagnetischer Felder (TEMF), 2010.

[133] M. Schwarz, L. Ewe, N. Hijazi, B. Hosticka, J. Huppertz, S. Kolnsberg, W. Mokwa, and H. Trieu, "Micro implantable visual prostheses," in *Microtechnologies in Medicine and Biology, 1st Annual International, Conference On. 2000.* IEEE, 2000, pp. 461–465.

[134] E. Shamonina, V. Kalinin, K. Ringhofer, and L. Solymar, "Magneto-inductive waveguide," *Electronics letters*, vol. 38, no. 8, pp. 371–373, 2002.

[135] ——, "Magnetoinductive waves in one, two, and three dimensions," *Journal of Applied Physics*, vol. 92, no. 10, pp. 6252–6261, 2002.

[136] Y. Shen and M. Z. Win, "Fundamental limits of wideband localization - Part I: A general framework," *Information Theory, IEEE Transactions on*, vol. 56, no. 10, pp. 4956–4980, 2010.

[137] Y. Shi, L. Xie, Y. T. Hou, and H. D. Sherali, "On renewable sensor networks with wireless energy transfer," in *INFOCOM, 2011 Proceedings IEEE.* IEEE, 2011, pp. 1350–1358.

[138] M. Sitti, H. Ceylan, W. Hu, J. Giltinan, M. Turan, S. Yim, and E. Diller, "Biomedical applications of untethered mobile milli/microrobots," *Proceedings of the IEEE*, vol. 103, no. 2, pp. 205–224, 2015.

[139] E. Slottke and A. Wittneben, "Accurate localization of passive sensors using multiple impedance measurements," in *Vehicular Technology Conference (VTC Spring), 2014 IEEE 79th*, May 2014, pp. 1–5.

[140] ——, "Circuit based near-field localization: calibration algorithms and experimental results," in *Vehicular Technology Conference (VTC Fall), 2015 IEEE 82nd*, September 2015.

[141] E. Slottke, M. Kuhn, A. Wittneben, H. Luecken, and C. Cartalemi, "UWB marine engine telemetry sensor networks: enabling reliable low-complexity communication," in *Vehicular Technology Conference (VTC Fall), 2015 IEEE 82nd.* IEEE, 2015, pp. 1–5.

[142] E. Slottke, R. Rolny, and A. Wittneben, "Universal computation with low-complexity wireless relay networks," in *Signals, Systems and Computers (ASILOMAR), 2012 Conference Record of the Forty Sixth Asilomar Conference on*, 2012, pp. 1770–1774.

[143] E. Slottke and A. Wittneben, "Single-anchor localization in inductively coupled sensor networks using passive relays and load switching," in *Signals, Systems and Computers (ASILOMAR), 2015 Conference Record of the Forty Ninth Asilomar Conference on.* IEEE, 2015.

[144] G. S. Smith, "Proximity effect in systems of parallel conductors," *Journal of Applied Physics*, vol. 43, no. 5, pp. 2196–2203, 1972.

[145] ——, "Radiation efficiency of electrically small multiturn loop antennas," *Antennas and Propagation, IEEE Transactions on*, vol. 20, no. 5, pp. 656–657, 1972.

[146] H. Sneesens and L. Vandendorpe, "Soft decode and forward improves cooperative communications," in *1st IEEE International Workshop on Computational Advances in Multi-Sensor Adaptive Processing, 2005.* IEEE, 2005, pp. 157–160.

[147] A. So and B. Liang, "Enhancing WLAN capacity by strategic placement of tetherless relay points," *Mobile Computing, IEEE Transactions on*, vol. 6, no. 5, pp. 522–535, 2007.

[148] J. J. Sojdehei, P. N. Wrathall, and D. F. Dinn, "Magneto-inductive (MI) communications," in *OCEANS, 2001. MTS/IEEE Conference and Exhibition*, vol. 1. IEEE, 2001, pp. 513–519.

[149] M. Soma, D. C. Galbraith, and R. L. White, "Radio-frequency coils in implantable devices: misalignment analysis and design procedure," *Biomedical Engineering, IEEE Transactions on*, vol. BME-34, no. 4, pp. 276–282, 1987.

[150] V. Stanford, "Pervasive computing goes the last hundred feet with RFID systems," *IEEE pervasive computing*, vol. 2, no. 2, pp. 9–14, 2003.

[151] C. Steiner, "Location fingerprinting for ultra-wideband systems: the key to efficient and robust localization," Ph.D. dissertation, ETH Zurich, 2010.

[152] C. Steiner, H. Luecken, T. Zasowski, F. Trösch, and A. Wittneben, "Ultra low power UWB modem design: experimental verification and performance evaluation," in *Union Radio Scientifique Internationale, URSI 2008*, Aug. 2008.

[153] C. J. Stevens, "Magnetoinductive waves and wireless power transfer," *Power Electronics, IEEE Transactions on*, vol. 30, no. 11, pp. 6182–6190, 2015.

[154] C. J. Stevens, C. W. Chan, K. Stamatis, and D. J. Edwards, "Magnetic metamaterials as 1-D data transfer channels: an application for magneto-inductive waves," *Microwave Theory and Techniques, IEEE Transactions on*, vol. 58, no. 5, pp. 1248–1256, 2010.

[155] I. Stojmenovic, "Position-based routing in ad hoc networks," *Communications Magazine, IEEE*, vol. 40, no. 7, pp. 128–134, 2002.

[156] Z. Sun and I. F. Akyildiz, "Magnetic induction communications for wireless underground sensor networks," *Antennas and Propagation, IEEE Transactions on*, vol. 58, no. 7, pp. 2426–2435, 2010.

[157] ——, "On capacity of magnetic induction-based wireless underground sensor networks," in *INFOCOM, 2012 Proceedings IEEE*. IEEE, 2012, pp. 370–378.

[158] M. Tamagnone, J. Gomez-Diaz, J. Mosig, and J. Perruisseau-Carrier, "Analysis and design of terahertz antennas based on plasmonic resonant graphene sheets," *Journal of Applied Physics*, vol. 112, no. 11, p. 114915, 2012.

[159] T. Theocharides, G. Link, N. Vijaykrishnan, M. Irwin, and V. Srikantam, "A generic reconfigurable neural network architecture implemented as a network on chip," in *Proceedings of IEEE International SOC Conference*, 2004, pp. 191–194.

[160] J. C. Tippet and R. A. Speciale, "A rigorous technique for measuring the scattering matrix of a multiport device with a 2-port network analyzer," *IEEE Transactions on Microwave Theory and Techniques*, vol. 30, no. 5, pp. 661–666, 1982.

[161] D. Tse and P. Viswanath, *Fundamentals of wireless communication*. Cambridge university press, 2005.

[162] Y. Tsividis, "Principles of operation and analysis of switched-capacitor circuits," *Proceedings of the IEEE*, vol. 71, no. 8, pp. 926–940, 1983.

[163] R. Twiss, "Nyquist's and Thevenin's theorems generalized for nonreciprocal linear networks," *Journal of Applied Physics*, vol. 26, no. 5, pp. 599–602, 1955.

[164] J. Unnikrishnan and M. Vetterli, "Sampling and reconstructing spatial fields using mobile sensors," in *Acoustics, Speech and Signal Processing (ICASSP), 2012 IEEE International Conference on*. IEEE, 2012, pp. 3789–3792.

[165] H. L. Van Trees, *Detection, estimation, and modulation theory*. John Wiley & Sons, 2004.

[166] Y. Wang, G. Leus, and A.-J. Van der Veen, "Cramér-Rao bound for range estimation," in *Acoustics, Speech and Signal Processing, 2009. ICASSP 2009. IEEE International Conference on.* IEEE, 2009, pp. 3301–3304.

[167] B. A. Warneke and K. S. Pister, "MEMS for distributed wireless sensor networks," in *Electronics, Circuits and Systems, 2002. 9th International Conference on*, vol. 1. IEEE, 2002, pp. 291–294.

[168] M. Wautelet, "Scaling laws in the macro-, micro-and nanoworlds," *European Journal of Physics*, vol. 22, no. 6, p. 601, 2001.

[169] W. Weitschies, J. Wedemeyer, R. Stehr, and L. Trahms, "Magnetic markers as a non-invasive tool to monitor gastrointestinal transit." *IEEE transactions on bio-medical engineering*, vol. 41, no. 2, pp. 192–195, 1994.

[170] W. Weitschies, M. Karaus, D. Cordini, L. Trahms, J. Breitkreutz, and W. Semmler, "Magnetic marker monitoring of disintegrating capsules," *European journal of pharmaceutical sciences*, vol. 13, no. 4, pp. 411–416, 2001.

[171] H. A. Wheeler, "Fundamental limitations of small antennas," *Proceedings of the IRE*, vol. 35, no. 12, pp. 1479–1484, 1947.

[172] H. A. Wheeler *et al.*, "Inductance formulas for circular and square coils," *Proceedings of the IEEE*, vol. 70, no. 12, pp. 1449–1450, 1982.

[173] A. Wittneben and B. Rankov, "Impact of cooperative relays on the capacity of rank-deficient MIMO channels," *Proceedings of the 12th IST Summit on Mobile and Wireless Communications*, vol. 2, pp. 421–425, 2003.

[174] ——, "Distributed antenna systems and linear relaying for gigabit MIMO wireless," in *Vehicular Technology Conference, 2004. VTC2004-Fall. 2004 IEEE 60th*, vol. 5. IEEE, 2004, pp. 3624–3630.

[175] L. Xie, Y. Shi, Y. T. Hou, and A. Lou, "Wireless power transfer and applications to sensor networks," *IEEE Wireless Communications*, vol. 20, no. 4, pp. 140–145, 2013.

[176] G.-Z. Yang and M. Yacoub, *Body sensor networks*, 2nd ed. Springer, 2014.

[177] L. A. Zadeh, "Time-varying networks, I," *Proceedings of the IRE*, vol. 49, no. 10, pp. 1488–1503, 1961.

[178] V. V. Zhirnov and R. K. Cavin III, *Microsystems for bioelectronics: scaling and performance limits*. William Andrew, 2015.

[179] L. Zhou, J. M. Kahn, and K. S. Pister, "Corner-cube retroreflectors based on structure-assisted assembly for free-space optical communication," *Microelectromechanical Systems, Journal of*, vol. 12, no. 3, pp. 233–242, 2003.

Curriculum Vitae

Name: **Eric Nathan Slottke**
Date of Birth: July 7, 1985
Place of Birth: Austin TX, USA

Education

07/2010-12/2016	**ETH Zurich, Switzerland** PhD studies at the Communication Technology Laboratory, Department of Information Technology and Electrical Engineering.
10/2004-05/2010	**Leibniz Universität Hannover, Germany** Studies in electrical engineering. Degree: Dipl.-Ing. (equivalent MSc.).
09/1996-07/2004	**Friedrich-Schiller-Schule, Leipzig, Germany** Abitur with focus on mathematics and physics.

Experience

07/2010-today	**ETH Zurich, Switzerland** Researcher at the Communication Technology Laboratory headed by Prof. Dr. Armin Wittneben.
10/2008-03/2009	**Volkswagen AG, Wolfsburg, Germany** Internship in the Group Research department.
02/2008-05/2009	**Institute of RF and Microwave Engineering** **Leibniz Universität Hannover, Germany** Student research assistant, Prof. Dr. Ilona Rolfes.
11/2006-10/2008	**Institute of Communications Technology** **Leibniz Universität Hannover, Germany** Student research assistant, Prof. Dr. Thomas Kaiser.

Publications

- **Conference, Symposium, and Workshop Papers**

    - **Magneto-inductive Passive Relaying in Arbitrarily Arranged Networks**
      G. Dumphart, E. Slottke, and A. Wittneben, IEEE International Conference on Communications (ICC 2017), Paris, France, May 2017.

    - **Single-Anchor Localization in Inductively Coupled Sensor Networks Using Passive Relays and Load Switching**
      E. Slottke and A. Wittneben, Asilomar Conference on Signals, Systems, and Computers, Pacific Grove, CA, USA, Nov. 2015.

    - **Circuit Based Near-Field Localization: Calibration Algorithms and Experimental Results**
      E. Slottke and A. Wittneben, IEEE Vehicular Technology Conference, VTC Fall, Boston, USA, Sept. 2015.

    - **UWB Marine Engine Telemetry Sensor Networks: Enabling Reliable Low-Complexity Communication**
      E. Slottke, M. Kuhn, A. Wittneben, H. Luecken, and C. Cartalemi, IEEE Vehicular Technology Conference, VTC Fall, Boston, USA, Sept. 2015.

    - **Accurate Localization of Passive Sensors Using Multiple Impedance Measurements**
      E. Slottke and A. Wittneben, IEEE Vehicular Technology Conference, VTC Spring, Seoul, Korea, May 2014.

    - **Universal Computation with Low-Complexity Wireless Relay Networks**
      E. Slottke, R. Rolny, and A. Wittneben, Asilomar Conference on Signals, Systems, and Computers, Pacific Grove, CA, USA, Nov. 2012.

- **Analysis of Channel Parameters for Different Antenna Configurations in Vehicular Environments**
  M. Schack, D. Kornek, E. Slottke, and T. Kürner, IEEE Vehicular Technology Conference, VTC Fall, Ottawa, Canada, Sept. 2010.
- **Constrained Maximum Likelihood Positioning for UWB Based Human Motion Tracking**
  Z. W. Mekonnen, E. Slottke, H. Luecken, C. Steiner, and A. Wittneben, International Conference on Indoor Positioning and Indoor Navigation, IPIN 2010, Zurich, Switzerland, pp. 1-10, Sept. 2010.
- **Effects of Antenna Characteristics and Placements on a Vehicle-to-Vehicle Channel Scenario**
  D. Kornek, M. Schack, E. Slottke, O. Klemp, I. Rolfes, and T. Kürner, IEEE International Conference on Communications (ICC 2010), Workshop on Vehicular Connectivity, Cape Town, South Africa, May 2010.
- **Experimental Investigation of Bent Patch Antennas on MID Substrate**
  D. Kornek, E. Slottke, C. Orlob, and I. Rolfes, 4th European Conference on Antennas and Propagation (EuCAP 2010), Barcelona, Spain, Apr. 2010.
- **A Simulation Study of Traffic Efficiency Improvement Based on Car-to-X Communication**
  H. Schumacher, C. Priemer, and E. Slottke, Proceedings of the 6th ACM International Workshop on VehiculAr Inter-NETworking (VANET), Beijing, China, Sept. 2009.

- **Talks**

  - **Inductive Coupling for Joint Power Supply and Communication at the Micro- and Nanoscale**
    E. Slottke and A. Wittneben, NaNoNetworking Summit (N3Summit), Barcelona, Spain, June 2012.

- **Patents**

  - **Method and System for Monitoring or Controlling an Engine**
    A. Wittneben, Z. W. Mekonnen, E. Slottke, M. Kuhn, and C. Cartalemi, European Patent Office, Patent application KR20160131914 (A), CN106124218 (A), Nov 2016.
  - **Method for Determining Relative Horizontal Angle Between Car and Transmitter for Car-to-X Communication Purposes, Involves Providing Evaluation Unit Evaluating Error Function, and Determining Angle Based on Values of Error Function**
    A. Kwoczek and E. Slottke, Patent application DE 102010019150 A1, Nov. 2011

Bisher erschienene Bände der Reihe

Series in Wireless Communications

ISSN 1611-2970

10	Stefan Berger	Coherent Cooperative Relaying in Low Mobility Wireless Multiuser Networks ISBN 978-3-8325-2536-1 42.00 EUR
11	Christoph Steiner	Location Fingerprinting for Ultra-Wideband Systems. The Key to Efficient and Robust Localization ISBN 978-3-8325-2567-5 36.00 EUR
12	Jian Zhao	Analysis and Design of Communication Techniques in Spectrally Efficient Wireless Relaying Systems ISBN 978-3-8325-2585-9 39.50 EUR
13	Azadeh Ettefagh	Cooperative WLAN Protocols for Multimedia Communication ISBN 978-3-8325-2774-7 37.50 EUR
14	Georgios Psaltopoulos	Affordable Nonlinear MIMO Systems ISBN 978-3-8325-2824-9 36.50 EUR
15	Jörg Wagner	Distributed Forwarding in Multiuser Multihop Wireless Networks ISBN 978-3-8325-3193-5 36.00 EUR
16	Etienne Auger	Wideband Multi-user Cooperative Networks: Theory and Measurements ISBN 978-3-8325-3209-3 38.50 EUR
17	Heinrich Lücken	Communication and Localization in UWB Sensor Networks. A Synergetic Approach ISBN 978-3-8325-3332-8 36.50 EUR
18	Raphael T. L. Rolny	Future Mobile Communication: From Cooperative Cells to the Post-Cellular Relay Carpet ISBN 978-3-8325-4229-0 42.00 EUR
19	Zemene Walle Mekonnen	Time of Arrival Based Infrastructureless Human Posture Capturing System ISBN 978-3-8325-4429-4 36.00 EUR
20	Eric Slottke	Inductively Coupled Microsensor Networks: Relay Enabled Cooperative Communication and Localization ISBN 978-3-8325-4438-6 38.00 EUR

Alle erschienenen Bücher können unter der angegebenen ISBN-Nummer direkt online (http://www.logos-verlag.de) oder per Fax (030 - 42 85 10 92) beim Logos Verlag Berlin bestellt werden.